UNDER THE WATCHFUL EYE

Under the Watchful Eye: Managing Presidential Campaigns in the Television Era grew out of a three-day symposium, "Campaigning for the Presidency," that was held December 5-7, 1991, at the University of California, San Diego. A public television special was aired about the symposium, which featured a wide-ranging discussion of campaign experiences and anecdotes. Participants, pictured above, included (from left to right) Gary Hart, national campaign director for George McGovern's 1972 presidential campaign; Susan Estrich, campaign manager for Michael S. Dukakis's 1988 campaign; Joseph Napolitan, director of advertising for the 1968 Hubert H. Humphrey campaign; Horace Busby, a speech writer for the 1964 Lyndon B. Johnson campaign; John Chancellor, commentator for NBC News and moderator of the televised symposium discussion; Robert Finch, Richard Nixon's national campaign manager in 1960 and an informal adviser to Nixon's 1968 campaign; Edward Rollins, national director of Reagan-Bush '84; Richard Kleindienst, national director of field operations for the Barry Goldwater for President Committee in 1964 and Nixon for President Committee in 1968; and Stuart Spencer, chairman of the 1976 Gerald Ford campaign and a campaign deputy for Ronald Reagan's 1980 presidential campaign.

UNDER THE WATCHFUL EYE

Managing Presidential Campaigns in the Television Era

Edited by

Mathew D. McCubbins
University of California, San Diego

A Division of Congressional Quarterly Inc.
Washington, D.C.

324.73
U55

Printed in the United States of America

Library of Congress Cataloging-in-Publication Data

Under the watchful eye : managing presidential campaigns in the
  television era / edited by Mathew D. McCubbins.
       p.   cm.
    Includes index.
    ISBN 0-87187-752-X
    1. Television in politics--United States.   2. Presidents--United
  States--Election.   I. McCubbins, Mathew D. (Mathew Daniel), 1956-

  HE8700.76.U6U53   92-24570
  324.7'3'0973--dc20   CIP

For Sue

CONTENTS

PREFACE

Nothing's real anymore unless it's on television.
—Mike Deaver, president,
Columbia Communications

Under the Watchful Eye: Managing Presidential Campaigns in the Television Era examines the changes that have occurred in presidential campaigns in the past three decades. The 1960 election was a last hurrah in American presidential election politics. The New Deal coalition—which was made up of city machines, unions, Catholics, blacks, Southern whites, and progressive reform intellectuals—held together for a final time. Voters still were loyal to their political party; party identification counted. Campaigns collaborated closely with local constituency groups and party organizations, which provided the bulk of campaign workers and funds. Nominations were made at the national party conventions, and the few primaries that were held mostly served to demonstrate a candidate's ability to attract voters, not to select the party's nominee. Since the watershed 1964 election, however, candidates have used primary campaigns to define themselves and their parties, and television has come to dominate the presidential election process. Conventions have become mere coronation ceremonies.

These changes have displaced the traditional means for selecting party nominees and presidents—by party organizations and by party bosses in smoke-filled rooms—with more candidate-centered processes. Some have argued that this transformation has led to the weakening of national parties and of the partisan links between members of Congress. Thus, changes in presidential elections have contributed to the collapse of the last vestiges of responsible party government at the national level. Government has become less responsive to the will of the people. Others have argued that the changes in presidential elections (including the central role taken over by television) have vastly increased voters' ability to make political (that is, voting) decisions without the mediation of the parties. The parties have been forced to become more attuned to the interests of voters, attending less to the more parochial interests of party bosses and machines. As a result, American government is more democratic and more representative.

Under the Watchful Eye grew out of a three-day symposium, "Campaigning for the Presidency," that was held December 5-7, 1991, at the University of California, San Diego, and broadcast nationally on PBS. Professionals with intimate knowledge of campaigns, past and present, were brought together to shed some light on, for example, the effects of the Democratic party reforms in the candidate selection process that were instituted in the late 1960s and early 1970s on such central issues as the quality of candidates and politicians' responsiveness to popular sentiment. However, according to the campaign experts, the impact of those and similar changes paled in comparison with the power of television. Media consultant Doug Bailey, for example, noted that "Judgments about presidential candidates now are made, for the most part, on the basis of the person that they see on television. . . . Therefore, withholding something relevant to the personal makeup of that person is more difficult and more questionable to do." In other words, voters are reacting to televised images, not partisan cues or policy positions. The focus of this book thus became the role of television in modern presidential campaigns.

Most of the credit for *Under the Watchful Eye* belongs to Sam Kernell, Robert Ritchie, and Mary Walshok. They were instrumental in organizing the symposium and getting the essayists to chair roundtable discussions and write their chapters. Special thanks goes to the Ford Motor Company, which generously underwrote the expenses of the symposium. All unattributed quotes in the pages that follow were taken from the roundtable discussions or the televised panel discussion that preceded them. John Aldrich wishes to thank Michael Alvarez and me for providing helpful comments and suggestions. He retains full responsibility for the content of his essay. Larry Sabato notes that portions of his chapter were drawn from his book *Feeding Frenzy: How Attack Journalism Has Transformed American Politics* (New York: The Free Press, 1991). He also wishes to thank University of Virginia graduate student Bruce Larson for his invaluable assistance in the design of the chapter, and Brian Sala, graduate student at the University of California, San Diego, for his help with revisions. Unless otherwise noted, all direct quotations in Sabato's essay were taken from the author's personal interviews with journalists and campaign aides or from comments made at the symposium. I thank Gary Jacobson, Sam Kernell, Rod Kiewiet, Skip Lupia, and Sam Popkin for their comments and editorial advice, as well as several anonymous reviewers for their comments and opinions on the entire manuscript. Lastly, I owe apologies to Joel Aberbach for appropriating the "watchful eye."

UNDER THE WATCHFUL EYE

INTRODUCTION

Mathew D. McCubbins

Academics and journalists of all ideological stripes are wont to complain that American democracy is in a shambles. Budgets don't balance, policies aren't responsive to the needs of citizens, incumbent members of Congress never lose, voters don't vote, and parties don't seem to matter. This sort of hand-wringing has become a popular American pastime, filling page after page of newspapers, magazines, books, and academic journals and trumpeted night after night on television newscasts.

Of the myriad perceived failures of our governmental system in the past several decades, the way we choose our presidents regularly tops the list. Americans (at least reporters and academics) love to talk about presidents and presidential elections. The American presidency seems to be a position of extraordinary power and influence in the world. As such, it provides a powerful focal point for public opinion about the state of the nation. Not surprisingly, when things are not going well for the nation, much blame is laid at the feet of the president, and much energy is expended in reexamining the process by which we choose our presidents.

The period since 1968 has been, in many respects, a tough one in the national history of the United States. The American economy had assumed such a predominant position in the world economy during the postwar years that we had perhaps come to expect as a birthright continued rapid growth in our standard of living, but the late 1960s and 1970s brought a series of debacles: the turmoil of Vietnam, the outrage of Watergate, the Arab oil embargo of 1974, the malaise of the stagflation of the mid-1970s, the Iran hostage crisis and second oil shock.[1] Each of these events shocked American sensibilities or pocketbooks. Add in the further economic and social upheavals rooted in the civil rights movement, the women's movement, and the coming of age of the baby-boom generation, and one can quickly see why Americans might have developed a sense of malaise about our nation's future and the ability of our presidents to right the ship of state.

New stylized interpretations of the presidential selection process first gained prominence in the 1970s. Whereas presidential campaigns were once said to have defined the national issue agenda and helped sweep majority parties into and out of power in Congress, today they seem to many observers to have less and less to do with real policy concerns and the governance of the American state, and more and more to do with photo opportunities and frivolous or sensational events: the sleeping habits of American politicians, the business partners of their spouses or nephews, or the criminal records of their brothers.

Presidential elections were not always like this: in the days when American democracy worked according to the Founders' design, giants seem to have roamed the earth. Abraham Lincoln, George Washington, Thomas Jefferson, and Theodore Roosevelt were the stock from which presidential timber was hewn. Even then, of course, not all presidents were destined to look down on the South Dakota badlands with stony visage for all eternity. But national crises such as the founding, the Civil War, and the rise of progressivism brought forward America's best.

The age of giants, many argue, passed with the demise of political party machines and the coming of the television age. These twin issues dominate the arguments about presidential campaigns in this volume. Each essay, at its heart, is about the dissemination of information and the coordination of voters' expectations and choices.

Before the development of television, according to the standard view, newspapers played a major role in presidential campaigns by disseminating information about primary candidates and coordinating voters by making endorsements. The print medium in the United States has always been intimately tied to political movements and political action, as is evidenced by the press's almost universal reverence for the First Amendment. In many cases, newspapers originated as arms of political organizations, run by and for members of political machines and their partisan identifiers.

However, the rise of television has, according to political consultant Doug Bailey, meant that "the whole [presidential] campaign is based upon what goes out on the nightly news on the television at night. That's what the campaign is." Americans today, it is claimed, get all or almost all of their political information from television. Unlike newspapers, however, television is not primarily an information service for consumers but rather an entertainment medium having more in common with cinema and works of fiction.

As a group, political reporters, broadcast or print, take very seriously their role as the "fourth estate"—sentinels for Americans' political freedoms. But, the story goes, the television medium and market make different demands on reporters than do print media and markets. This structural difference has had important consequences for television news coverage, for, as former White House adviser Michael Deaver said,

> One of the real problems [with television political reporting] that television simply will not come to terms with, is that they are an entertainment medium. They are not a news medium. Particularly, when [viewers] are sitting there with their clickers and they've got 69 channels to go around, you've got to stop that clicker, and you've got a second to do it. . . . The problem for us—when I say "us" I'm talking about this society—is that, unfortunately, that's where most of us are getting our information, with that clicker.

Television relies first and foremost on the images it presents to the viewer, not just the quality and quantity of the information in an encyclopedic sense of facts and figures, but stylistic qualities and emotive capacity that seem to be lacking in other media. It allows candidates to step into every American's living room, to look him or her in the eye and make an immediate, emotional, gut impression. Critics argue that the instant emotional gratification that television can offer viewers conditions the kinds of competition that news reportage must face, and hence the kinds of reportage likely to be aired.

It is often argued that television has increasingly slighted campaign issues in favor of catchy imagery and soundbites. Recent research by Dan Hallin[2] shows that television coverage of candidate press conferences has moved away from the longer audiovisual clips often used on the nightly news shows in the 1960s toward the very brief soundbite clips used today, annotated by the reporter's interpretation or analysis. We are less able to read the lips of our candidates today because TV shows many fewer candidates—and many more reporters.

This change in coverage, it is argued, is emblematic of the increasing gap between the kinds of reporting done by television and newspapers. Critics of television reporting observe that there is less straight presentation of news events and more interpretation being done by TV reporters. These critics argue that TV newscasts now play a much more active role in shaping and defining what the news means for viewers. This changed mission, in turn, is reflected in the different approaches that newspaper and TV reporters take to researching stories. TV reporters, it is argued, have turned reporting upside down. They now look for the catchy image or soundbite around which to build a story instead of building a story and then writing a lead.

This preoccupation with imagery, it is argued, fails to provide voters with the kinds of information they need to make informed choices between presidential candidates. It provides pictures that TV thinks will grab viewers, not information relevant for evaluating the candidates' issue positions or competence. Even more disconcerting to critics of television is its indirect effect: the emphasis on images changes the types of people who think they can win elections, encouraging those best able to "sell" a good image while discouraging those whose attributes don't translate well to the small screen and facile "bites."

A further consequence of the new reliance on television as the primary medium through which the public comes to know presidential candidates is a phenomenon known as momentum: the tendency for campaign or primary successes to build on themselves, or, alternatively, for campaign gaffes or failures to snowball. John Aldrich argues in an essay in this volume that the television medium has fundamentally changed the way presidential campaigns are organized and run in order to try to keep the candidate in command of his own momentum. Gary Hart's story is just one example of the momentum phenomenon. An unexpected groundswell of support for Hart in New Hampshire in 1984 transformed a relative unknown into a serious challenger overnight; likewise, sudden revelations about a cruise to Bimini transformed him from nominee-presumptive to ex-candidate in a matter of days in 1988. The challenge for campaign organizations today, Aldrich argues, is managing the interplay between mass opinion and events rather than stitching together organized groups into coalitions. Thus, where primaries and caucuses were once most notable for allocating convention delegates, today they are also used by candidates and their "spin doctors" to shape the media's expectations and voters' choices in subsequent states.

Momentum comes down to a coordination of voters' expectations that a particular candidate is most likely to win the nomination. Television, it is argued, radically compresses and accelerates the process of transferring information from candidate to audience, allowing a candidate to influence mass opinion directly rather than through the mediation of organized groups. This makes it increasingly difficult for relatively unknown, moderate candidates to build a coalition of support: well-known candidates have a tremendous initial advantage in gaining the attention of TV, since they are known commodities with known appeal to viewers. Ideologically extreme candidates, on the other hand, have an advantage over centrist candidates with similar name recognition: they have novel, perhaps inflammatory messages that often make for dramatic imagery. Thus in a competition for the attention of viewers, voters, and potential contributors at the start of campaigns, centrist candidates can get squeezed out, much as reliable but dull products get squeezed off supermarket shelves by the competition from market leaders and new products.

The competition for air time is, in effect, a contest for which candidates will get on the agenda—which candidates voters will even consider when they finally cast their ballots. Critics of TV's effects on campaigns argue that this competition, and the subsequent process of momentum shifting with the wind currents of events, distances presidential campaigns from any real debate over issues. Hence, television, by putting vast quantities of imagery-based information in the hands of viewers, has actually promoted voter ignorance.

The consequence of increased voter ignorance, implicit in the standard view, is that electoral outcomes (both in presidential nominations and in general elections) are more volatile. The momentum story implies that the choice of the marginally attentive average voter is made for him or her by the media, while the attentive voter is unable to collect enough information from candidate imagery to make an informed choice. As a result, these arguments imply, attentive voters are much less likely to coordinate on a "best" candidate to outweigh the effects of average voters who respond to images. Bad candidates are more likely to be chosen and therefore bad presidents are more likely to be the result.

In the olden, golden days before television, these critics imply, party bosses were able to select electable, competent candidates to run for office and could forge a unified party from the factional, regional splits, i.e., assemble a national coalition of voters. With the decline of party, Madison Avenue and Hollywood have supplanted party bosses in the presidential selection process. The successful candidate is defined more by the ability to attract marketing wizards and dollars at Hollywood parties than by competence or issue positions. The parties are no longer parties; they are houses divided, standing only by the dint of ossified party identification in the herd of tube-tied voters.

In this book, John H. Aldrich, F. Christopher Arterton, Samuel L. Popkin, Larry J. Sabato, and I reexamine the claims that television and party decline have changed presidential campaigns, the evidence that presidential campaigns have in fact changed, and the effect of these changes on American democracy.

The plan of the book is simple. In the first chapter, I present a theoretical overview of the literature on presidential campaigns. Laying out the predominant view of how presidential campaigns have evolved, from this stylized view I construct an internally consistent, parsimonious explanation for both pre-TV and modern presidential campaigns. Then, I survey the evidence offered in support of the predominant view and attempt to reconcile this evidence with expectations derived from the explanation presented, rejecting those claims unsupported by the evidence or reasonable argument.

What's left is a much smaller set of stylizations of presidential campaigns. First, evidence and theory support the notion that certain changes in state electoral laws and party rules have affected presidential campaigning and elections. Changes in state ballot laws before the turn of the century and the move to primary elections for many important elective offices led to the demise of party machines in the United States. Second, these legal changes also partly explain the rise of candidate-centered campaigning in this century. Third, it is clear that television has changed some aspects of American life. However, I find no evidence nor any good theoretical reason to believe that TV has made any real difference in presidential elections. Television has not reduced the ability of the print media to deliver the news, nor has it created a nation of illiterates and ignoramuses (despite the evidence of real problems with our education system). If anything, TV has increased the amount of political information available to the average person in America.

What I do draw from the literature on TV and politics is that most scholars have thought wrongly about what "information" is and what role it plays in decision making. Information is not merely facts and figures, i.e., encyclopedic data. Information is endorsements, images, "narratives"—any environmental signal received by a person that can help that person choose between alternative courses of action. What matters in decision making is not how many facts a person knows but rather how accurately he or she is able to make inferences about the consequences of his or her choices.

The chapters that follow my opening essay take up the issues of candidate-centered campaigns and momentum, the marketing of presidential candidates through TV advertising, the narrowness of modern political reporting, and the ability of voters to extract value from the political imagery they receive. Each essay wrestles with an interesting aspect of how modern presidential campaigns differ in style and structure from pre-TV contests. These essays are valuable, not because they contradict my findings, for they do not, but because they clearly illustrate the ways in which modern campaigns are different from their predecessors.

I have already mentioned the focus of John Aldrich's essay, which deals with the rise of candidate-centered campaign organizations. His account revolves around the issue of how the momentum phenomenon, bred from weak parties and the rise of television, has changed the competitive pressures candidates face. The decline of party machines, he argues, made candidate-centered campaigns possible; television extended the range of such campaigns to presidential contests. The crux of Aldrich's argument is that modern campaign organizations are event-driven rather than issue-driven: candidates rush back and forth to take advantage of good press reports and squelch bad ones about their records and personal lives instead of concentrating on refining and advertising the substantive policy positions each is trying to sell. Thus, for candidates in the TV age, image is everything.

In the third essay, Christopher Arterton challenges the conventional view (and the implication of Aldrich's essay) that the message component of campaigns does not matter. He argues, however, that this component is biased by the television medium as campaigns try to tailor their message to attract undecided voters. Candidates start from a base of loyal supporters and attempt to add a sufficient number of ideologically more distant voters in order to win the nomination or election. But, Arterton argues, too much attention to attracting these marginal voters has tended to alienate voters in the base; the general consequence of television competition has thus been a shallowing of support for all candidates and a continuing process of dealignment in the electorate, as voters no longer see good reasons to adhere to any candidate or party.

In the fourth essay, Larry Sabato extends arguments about the alienating effects of modern media coverage on prospective voters. He argues that structural changes in media markets and a loosening of libel law following *The New York Times v. Sullivan*[3] in 1964 led to a new equilibrium in news coverage; political coverage has become more sensationalist, entertainment-oriented, and frenzied, in the sense that *stories* develop momentum. Sabato thus provides one potential underpinning for the political momentum phenomenon identified by Aldrich.

The broad, antidemocratic implications of television's impact on presidential campaigns, offered by Aldrich, Arterton, and Sabato, are challenged by Sam Popkin in his essay. He maintains that voters are much more sophisticated image processors than is commonly thought. What matters is not the encyclopedic information in a particular advertisement, endorsement, or image consumed by a TV-viewing prospective voter but rather how the image or endorsement conforms to a network of prior observations and experiences the voter has already internalized. Popkin argues that what matters most are the cognitive models that voters employ to interpret the images and signals they consume; more powerful models allow more accurate inferences about the likely consequences of electing one candidate over another. Hence, the vital question for present-day presidential campaigns is not what information content is provided by TV advertisements or news in comparison with what advertising and media used to provide, but whether voters' cognitive models are sophisticated enough to keep up as TV ads and news have become more image-oriented. Popkin's answer is, in brief, that *campaigns*, not voters, have fallen short and that we now need bigger, louder, longer presidential campaigns that can provide the volume of information that voters are capable of processing. The volume concludes with an essay in which I summarize the arguments and evidence presented by Aldrich, Arterton, Sabato, and Popkin in order to address the dreaded "so what?" question ubiquitous to social scientific research. Which of the hypothesized changes in presidential campaigns have actually changed since the 1960s, and what consequences have those changes had for the democratic responsiveness and efficiency of American political institutions, particularly the presidency? Specifically, I examine the McGovern-Fraser and later reforms, since much blame has been laid at the feet of these changes in nomination rules and procedures.

Notes

1. For a discussion of the socioeconomic issues underlying the stylized interpretations of the state of the economy in the 1970s and 1980s, see Frank Levy, *Dollars and Dreams: The Changing American Income Distribution* (New York: Russell Sage Foundation, 1987).
2. Daniel C. Hallin, "Soundbite News: Television Coverage of Elections, 1968-88." Occasional paper, Woodrow Wilson International Center, Washington, D.C., 1991.
3. *The New York Times v. Sullivan*, 376 U.S. 254 (1964).

1. PARTY DECLINE AND PRESIDENTIAL CAMPAIGNS IN THE TELEVISION AGE

Mathew D. McCubbins

Scholars and popular observers alike are in nearly universal agreement that the decline of parties and the rise of television have had a great impact on American democracy. In this chapter I reassess the claims made about these two "revolutions" in American politics and their consequences for our democracy.

I begin by discussing the notion that a golden age of parties and party government preceded the modern period and the advent of TV. I then draw a set of stylized depictions of modern American politics from the scholarly literature on presidential campaigns and present the commonly cited consequences for American democracy of these changes wrought by TV and party decline.

It is insufficient merely to list the details of this revolution in presidential campaigning. If we truly believe that American democracy is on the road to ruin, we cannot hope to halt the process of decline without understanding the roots of that process. Thus, I offer stylized arguments to explain the changes in presidential campaigns. The presentation of these stylizations raises numerous questions as to how factual the stylized facts really are. I next examine the evidence for and against these stylizations. I then conclude by focusing attention on those depictions of modern elections for which some basis in reality can be shown by means of plausible evidence or reasonable argument.

The gist of my argument is that although party competition and campaigning look a lot different today than they did fifty or a hundred years ago, neither the changes in parties and campaign organizations, nor the change in information transfer brought about by television, have fundamentally altered the outcomes produced by presidential campaigns. The presidents that we select today are not systematically less responsive to centrist ideological interests than presidents in the nineteenth century, nor are they systematically less competent, more corrupt, more "imperialistic" toward Congress, more adept at building legislative coalitions, or more photogenic. Indeed, if theory and evidence do indicate any change, it is that presidential candidates today are more candid about their records and more concerned about being responsive to voters' interests than ever before.

I do not argue against the idea that modern presidential campaigns are candidate-centered, in the sense that much of the news coverage and campaign action now seems to revolve around the personal characteristics of the candidates. This focus is the result of electoral laws and rules that destroyed party machines, beginning in the 1880s, and does not imply anything about the electoral value of a candidate's party label. Continuing evidence of the importance of those labels is the fact that no third-party candidate has outpolled either the Democratic or the Republican candidate in November since Theodore Roosevelt finished second in 1912. Indeed, only a handful of presidential candidates in this century have won any electoral-college votes without running under a major party label (Roosevelt, Robert La Follette, Strom Thurmond, and George Wallace being the most prominent, each of whom led a splinter faction from one of the major parties).

Nor do I find any evidence or reason to believe that TV has made any real difference in presidential elections in terms of the kinds of individuals who are chosen or the kinds of policies they pursue in office. What I do find in the literature is that scholars have not only misunderstood how and why people gather information to make decisions but also what counts as information.

Finally, I have found no evidence or reason to believe that either TV or candidate-centered campaigning has been bad for American democracy. The hand-wringing about the state of American democracy, which is so evident in the popular press every four years, has more to do with divided government than with bad presidents. Scholars and the media have misinterpreted the role of the president in our constitutional system, aggrandizing what Wilson called our "chief clerk" into a latter-day caesar. Our president is clearly an important player in the legislative process—the veto power assures him of that role—but he is not, and was not meant to be, our Kwisatz Haderach. To attribute god-emperor status to the office is to ensure that its occupants will fall short of expectations.

When Giants Roamed the Earth: Presidential Elections before Television

There is widespread agreement in the literature on five general characterizations of twentieth-century presidential electioneering before the popularization of television in the 1960s. First, *despite losing control over voters, a coalition of state and local party machines across the nation maintained control over presidential nominations.* Nelson

Polsby and Aaron Wildavsky characterized national parties in this era as "coalitions of state parties which meet every four years for the purpose of finding a man and forging a coalition of interests sufficiently broad to win a majority of electoral votes." [1] In characterizing conventions of this era, Polsby and Wildavsky argued:

> A relatively few party leaders control the decisions of a large proportion of the delegates to conventions. Delegates to national conventions are chosen, after all, as representatives of the several state party organizations. . . . While it is true that official decisions are made by a majority vote of delegates, American party organizations are centralized at state and local levels. This means that such hierarchical controls as actually exist on the state and local level will assert themselves in the national convention. [2]

Party bosses had to be able to deliver a bloc of votes at election time if they were to wield any influence over the nominating process, even though party rules may have concentrated the authority to name convention delegates in the hands of party bosses. Machine endorsements of candidates could have a powerful coordinating effect on voter choice: by committing the votes that it controlled to a particular candidate in a multicandidate setting, such as a primary, the machine could tag its favorite as the front-runner. Rival candidates effectively faced a barrier to entry; how were they to coordinate enough voters to neutralize the bloc of voters committed to the machine candidate? Voters, who presumably had little information on which to base their choices, were forced to choose between a relatively clear choice (the machine-endorsed candidate) and a grab bag of lesser-known alternatives. Very often voters followed the cues provided by political machines. [3] But the grasp of political machines on the reins of political power loosened over a period of several decades after the turn of the century. It is thus difficult for the literature to pinpoint exactly when machines no longer controlled enough votes to dominate the nomination process. [4]

Second, *machine bosses chose electable candidates who could lead the ticket to victory.* Theodore White noted:

> Big-city bosses can usually be swayed only by a national candidate's demonstration of surplus disposable power at the polls, by a political glamor demonstrated as so appealing that voters will indiscriminately in November fatten the local ticket at the base and strengthen the machine's local candidates where it counts, at home. [5]

One way for the bosses to gauge a potential nominee's expected surplus was to observe the success of candidates in those states that held primary elections, although, as Polsby observes, "winning even a large proportion of them was not a sufficient condition for winning the nomination." [6] Thus, in order for candidates such as John F. Kennedy to win the confidence and support of the party bosses, White noted, "there was no other than the primary way to the Convention. If they could not at the primaries prove their strength in the hearts of Americans, the Party bosses would cut their hearts out in the back rooms of Los Angeles." [7]

John Geer argues, however, that presidential primaries between 1912 and 1968 served largely an advisory role to party bosses and were not the sole determinants of who would bear the party banner in November. "[A] series of victories in the primaries did not guarantee anything. For instance, in competitive races for the nomination, the top vote getter in the primaries became the nominee only about 40 percent of the time, showing that party leaders often ignored the advice offered by voters in primaries." [8]

Third, *machine bosses chose competent candidates who were qualified to be president.* Party leaders and bosses thus acted as a filter to weed out amateurs and incompetents in the presidential selection process. According to Polsby, "[T]he criterion of peer review seeks to increase the confidence of voters in the capacities of the nominee actually to execute the office of the Presidency effectively." [9]

Party bosses did not perform this function out of the goodness of their hearts or love of country, however. Basically, according to Polsby and Wildavsky, the party leaders were willing to do what was necessary "to gain power, to nominate a man who can win the election, to unify the party, to obtain some claim on the nominee, to protect their central core of policy preferences, and to strengthen their state party organizations." [10]

Fourth, *machine bosses chose their party's platform.* Party platforms were chosen that would appeal to voters, and often, as Martin Wattenberg notes, "parties forced presidential candidates to moderate their views and move toward the center." [11] In summation, the party bosses—the very same people who were decried by Progressives and municipal reformers for their nefarious effects on state and local politics—have been lauded for performing a host of salutary functions vis-à-vis presidential selection. Responsible national party government, according to this standard view, depended to a great extent on the self-interest of local party bosses.

Fifth, *newspapers were local in scope and often overtly political and partisan.* As Lance Bennett explains, "Reporting involved the political interpretation of events. People bought a newspaper knowing what its political perspective was, and knowing that political events would be filtered through that perspective." [12] There was no pretense that information—what there was of it—provided by the media was unbiased or apolitical. Newspapers were either the tools of political machines or their rivals for political influence. As Robert Finch put it, there was a time "when the Copley papers and the Nolan papers and the *L.A. Times* and the Hearst papers would mark a ballot from top to bottom and usually decide primary elections, which they did, with a marked ballot."

The old world of presidential elections, then, was characterized in the literature by three factors: voters had access to only a small set of highly biased yet consistent information sources; presidential nominations were controlled by a handful of party bosses whose political interests were fundamentally local, not national; and these bosses, following their self-interest, limited nominations and party platforms to candidates and issues that were responsive to broad national interests in order to maximize the positive coattails that a party presidential nominee would provide for the machines' local tickets.

Under the Watchful Eye: Presidential Campaigns in the Television Age

The literature also shows widespread popular agreement on a number of characterizations of modern presidential campaigning. These also relate to the role and function of party organizations and the media in the political process, but in contrast to the preceding age, the modern era is said to be marked by trends toward the nationalization of information and the devolution of party power to the mass media. Much of the literature (exclusive of the authors in this volume) argues that although the machine-dominated era was not particularly democratic, its passing has led to a poorer class of candidates and presidents. Second, it is generally argued that many of the political and educative functions once fulfilled by political machines and partisan newspapers have now fallen to television, whose status as an entertainment medium critically damages its ability to serve those functions adequately. American democracy has suffered as a result.

The remainder of this section is a point-by-point explication of these two basic claims. As in the first section, here I sketch out the popular characterizations in order to present a robust picture of the current dominant view of the presidential selection process.

(1) It is universally argued that *the power of party machines to control votes, and therefore the nomination, has disappeared.* As White described it,

By 1968 [the] old political map of the Democratic party was as out of date as a Ptolemaic chart of the Mediterranean. No one any longer controlled New York, the state which had given Lyndon Johnson a 2,669,534 plurality in 1964. When one telephoned a switchboard of the New York State Democratic Committee early in 1968, one heard the languid voice of the telephone girl, "No. He's not here. There's nobody here. No one comes in any more. You can leave a message if you want." . . . Here and there, of course, a few hereditary enclaves of the old machines, passed on from father to son, still existed. . . . But, over all, the machinery creaked and clanked, as if it had come ungeared.[13]

The political machines, after decades of decline, had finally become so feeble that their leaders could no longer dominate the conventions or primary elections. They were no longer able to forge a national coalition from the various interests and factions that make up the national parties.[14] This, of course, implies that the *guiding force of the party bosses in choosing electable and competent candidates was lost as well.* This point will be raised again in the context of our discussion of the rise of the modern media.

The state and local machines also had provided ready-made organizations with resources and manpower needed to fuel a national campaign. Without the political machines to act as coalition builders and providers of campaign labor and expertise, prospective presidential candidates were forced to turn elsewhere. Just as congressional candidates before them had done in the wake of state-sponsored congressional primaries decades earlier, presidential candidates increasingly turned to creating their own organizations.

(2) It follows that *successful candidates have their own large and expensive campaign organizations.* Winning candidates are likely to be those who succeeded in building organizations comparable in vote-getting capacity to the old networks of machine organizations. Barring changes in technology, those organizations would have to be comparable in size to the combined magnitude of the machine networks. And unless those organizations were completely dependent on volunteer labor and/or employees on contingency contracts (such as quid pro quo expectations of job offers in the new president's administration), they were bound to be expensive as well. It is clear that the total cost of presidential nominations, for both parties, has climbed considerably in recent decades. In 1968 the Democrats and Republicans spent a total of $129 million in 1982-1984 dollars, while in 1988 the two parties spent a total of $174 million (in constant dollars).[15]

Presidential campaigns have become not only increasingly expensive but also considerably longer. Howard Reiter shows that major candidates in contested Democratic conventions have been formally announcing their candidacies much earlier than in the past; the number of days before the party convention has risen from an average of 148.5 days in 1932 to 427.3 days in 1984. Similarly, for the Republicans, it has risen from 303.3 days in 1940 to 337.5 in 1976.[16]

Polsby argues that the increased use of primaries, and the concomitant need for ample funds to compete in a long national campaign,

> have given rise to the displacement of state party leaders and leaders of interest groups associated with them in the presidential nomination process. ... It has ... meant that a new group of political decision-makers has gained significant authority. These are the fund-raisers by mail and by rock concert, media buyers, advertising experts, public relations specialists, poll analysts, television spot producers, accountants and lawyers who contract themselves out to become temporary ... members of the entourages of presidential aspirants. They work not for the party but directly for the candidates.[17]

Hence, *consultants have replaced party leaders in key campaign roles.*[18] For Doug Bailey, this problem goes beyond just campaign staff; modern presidential candidates are

> both proud and willing to accept the consulting mechanism that seems to come with the campaign. They're willing to subordinate themselves to the handlers.
>
> And if that's the type of person ... that we're attracting as candidates then we've got a serious problem, because ... the consultants are ahead of the candidates—it's not the candidate's campaign any more.

Thus the role of party and interest group leaders in presidential campaigns has changed from one of primacy to one of advocacy. Today, as Polsby points out, "selected state leaders may be invited to participate in a candidate's campaign rather than the candidate being a recruit and representative of the state party or faction."[19] Party leaders, it is argued, have gone from selecting candidates and running their campaigns to having almost no role in presidential elections. As Asher notes, "The presidential contest today is dominated by teams of specialists and loyalists whose first obligation is to the candidate and not to the political party of which that candidate is a leader."[20]

An unexpected—though perhaps predictable—consequence of this shift to *candidate-centered campaigns* has been that the issues and discussions now focus on the candidates and their personal qualifications and character rather than primarily on party platforms and party reputation. As Polsby and Wildavsky explain, "Since candidates self-select and campaign for themselves through the media, it would not be surprising if the electorate, observing what has happened, pays more attention to the prime movers, the candidates, and less to their party labels." [21]

(3) *Voter participation and party identification have declined.* Following the institution of the Australian ballot, voter registration laws, and other ballot reforms around the turn of the century, the rate of participation of eligible voters in presidential elections declined from approximately 80 percent, for the period from Reconstruction through 1896, to 59 percent by 1912.

In support of the statement that party identification has declined, two arguments are usually employed. First, survey research has indicated that the percentage of voters who identify themselves with the Democrats or Republicans has declined, while those who identify with neither of the major parties has risen over time. Wattenberg presents National Election Study data for presidential years 1964 through 1980, breaking down the "nonidentifiers" into apoliticals (1 to 2 percent of eligible voters polled), no-preference (rising from 2 percent of those polled in 1964 to about 10 percent in 1980; about half of this group consistently admits to "leaning" toward one or the other party, however), and independents (averaging about 25 percent of those polled, two-thirds of whom reported "leaning" toward one or the other party).[22]

Second, it is argued that the degree of split-ticket voting has increased dramatically during this century, robbing survey measures of party identification of much of their value. Reiter asserts that the increase, even in recent years, has been significant:

> From 1952 to 1964, some 11 percent to 17 percent of respondents split their tickets in races for the Presidency and the Senate (not counting votes cast for third-party and independent candidates). In 1968 that figure rose to 19 percent and has exceeded 20 percent ever since. Similarly, in voting for President and Representative, the figure rose from 12 percent to 16 percent from 1952 to 1964, to 18 percent in 1968, and at least 25 percent ever since.[23]

(4) It is argued that *nominations are more contestable than they were in the era of strong parties and bosses.* People have two developments in mind: first, that there are more candidates contesting nominations than in the past. Polsby tells us that the number of Democratic candidates running in more than one primary increased from a mean of 3.2 in 1952-1968 (before the nomination reforms) to 13 in both 1972 and 1976.[24] The Republican side also saw an increase, from the same 3.2 mean in 1952-1968 to 9 candidates seeking to oust Carter in 1980.

The second claim, as Stephen Wayne states, is that "opportunities for outsiders have increased." [25] Before the decline of party and rise of television, most politicians wanting to run for the presidency had no hope of winning, whereas now many more prospective candidates foresee a positive (if still small) chance of winning their party's nomination. It follows from the greater openness of a primary-focused nomination process that *amateurs, opinion outliers, and interest groups can under the right circumstances take over the nomination process* as well as write the party's platform.

In 1979, for example, the National Education Association (NEA) officially supported Carter's renomination campaign, endorsing him when his poll standings were particularly low. Teachers worked hard to support Carter, and in the summer of 1980 the NEA supplied 302 delegates to the Democratic convention in New York. They received financial support from the NEA, caucused on every issue before the convention, and were monitored in floor voting by NEA whips. This sort of machine-like activity helped defeat Edward Kennedy's convention challenge and gave the disciplined cadres of the NEA a strong voice in the writing of the Democratic platform. [26]

Jeane Kirkpatrick presents evidence that delegates to the 1972 Democratic convention held opinions far different from Democratic identifiers in the electorate. [27] Everett Carll Ladd goes on to provide figures for the 1976 Democratic convention suggesting that although the delegates were again far different from Democratic identifiers in the electorate, they were far closer in their opinions to "New Class" Democrats—college-educated professionals under forty years old. (Ladd wrote this book before "yuppie" entered the popular idiom.) The "New Class" Democrats

> support the busing of school children to achieve racial integration; ... reject "equality of opportunity," insisting instead upon "equality of result"; ... want to extend civil liberties, notably the rights of the accused in criminal trials; and [who] sharply question the value of economic growth, believing that it damages the "quality of life." The New Liberalism also differs from the New Deal ethos in the matter of personal morality; it takes a libertarian stance on such issues as abortion, legalization of marijuana, homosexuality, and racial intermarriage. [28]

When such unrepresentative convention delegates do gain control of their party's convention, *the result is likely to be the nomination of an outlier candidate.* According to Polsby and Wildavsky, "sometimes the policy positions advocated by party activists tend to be unpopular with most other people. Thus, it is possible for the nominating process to produce candidates who appeal to the people who become delegates but not to voters." [29] Richard Watson suggests:

> The increased influence of amateurs dedicated to issue-oriented candidates means that both parties run the risk of nominating candidates whose views on public policy do not correlate with views of rank-and-file voters. When they do—as the Republicans did when they nominated Barry Goldwater in 1964 and Democrats did when they chose George McGovern in 1972—the result is the mass defection of traditional party supporters and the overwhelming defeat of the party's candidate in the general election. . . . Another serious possibility is that the amateurs *in both parties* could be successful in the same year and confront the electorate in the November election with having to make a decision between a Goldwater and a McGovern. [30]

The McGovern-Fraser reforms (and similar reforms undertaken in the Republican party) are thought to have enhanced these trends. [31] The McGovern-Fraser Commission on Party Structure and Delegate Selection produced a set of guidelines that it claimed constituted "the minimum action state Parties must take to meet the requirements of the Call of the 1972 Convention." [32] The guidelines ruled out the use of party caucuses and "delegate primaries" to select delegates in favor of more open conventions and "candidate primaries," in which prospective delegates would run as committed to a particular presidential candidate. They also called for greater demographic representation in state delegate slates and closer conformity between a candidate's relative support in a state and the number of delegates he or she received. As a partial response to the McGovern-Fraser Commission, the number of states with Democratic primaries rose from seventeen in 1968 to twenty-three in 1972, while the proportion of convention delegates selected in primaries rose from about 50 percent to two-thirds. Again, the Republicans undertook similar changes in the number of primaries and proportion of delegates selected in primaries.

(5) It is argued that *voters get most of their information from television* and that television coverage of campaigns shapes how the rest of the media reports on candidates as well as how candidates and their consultants shape their campaign strategies. [33] According to *Los Angeles Times* reporter George Skelton, numerous polls conducted by the newspaper support this claim. "In fact, every poll we do, we ask the question, 'Where do you get most of your information?' And over a long period of time, many polls, ordinary people [say they] get most of their information from television." The central role of television in information dissemination in the United States is widely believed. Mike Deaver offered a very typical view: "80 or 90 percent of the people in

this country get all their information and make all the decisions about the way we run this country from television."

(6) Critics charge that *the media in general—and television in particular—ignore questions of candidate competence and policy in their coverage of campaigns.* For Gary Hart, this tendency has been both clear and unfortunate. Referring to the 1984 campaign for the Democratic nomination, Hart tells of a "serious speech on foreign policy" that he gave in Chicago:

> Mondale was charging me with something about an ad in the Chicago and Illinois primary, and I was responding, unfortunately, and it was back and forth. I'll never forget the front page of the *Chicago Sun-Times*: "Hart responds to Mondale charge," paragraph, paragraph, paragraph, paragraph, paragraph, paragraph. And the last sentence of the story is, "Hart was in Chicago to give a major foreign policy speech." Now, from that newspaper, you would never have known what the speech was about. It's politics. It's politics, politics, politics, politics. And I don't think people want to hear that.

While this anecdote refers to newspaper coverage, it applies in spades to TV. Television, it is argued, is first and foremost an entertainment medium, not a news medium. As Doug Bailey notes, viewers can switch channels merely by pressing a button on the remote control. "The clicker means you can do it without getting up out of your chair, right? So it's very easy to change channels. The battle for getting your attention becomes greater and greater; therefore, more excitement." For George Skelton, this seems natural because "it's more fun. We are in this business to tell the reader about a competition. It's a struggle for power that's going on in the country. People want to know constantly who's ahead, who's doing what to who to get there, what strategy is being used. I mean, that's why we read the sports page."

The clicker and the nature of television as an entertainment medium, according to Doug Bailey, explain a major difference between television and print: where print reporters need to fill up space and get a story—not just a single quote—a television interview will continue only "until that reporter has heard the soundbite, and then the interview is over. I mean, that's what it's all about. It really doesn't have very much to do with analysis at all." Thomas Marshall states the case succinctly: the way that the media cover primaries means that "voters are provided little issue information—but much information on campaign trivia and on the candidates' win-loss records." [34] Robert DiClerico and Eric Uslaner show that news coverage of primary campaigns in 1976 dealt with issues only 24 to 28 percent of the time; the rest of the coverage was devoted to "horse race and hoopla." [35]

The success of TV has affected all reporters, prompting the growth of what Sabato has termed "junkyard dog," or attack journalism.[36] Attack journalism (or a journalistic "feeding frenzy") can be seen as the consequence of two tendencies in the journalistic community: first, a tacit coordination mechanism that induces homogeneity in how stories are framed—pack journalism. Second is the competitive pressure of the marketplace. Newspaper editors are driven to a great extent by the twin desires to scoop the competition and to avoid being scooped. This means simultaneously covering everything the other guy does, plus perhaps a little more. According to Jerry Warren, "The competitive urge in Washington [in the 1988 campaign] . . . was so great that they would go after anything, and so there were a number of cases where rumors were put on the wire and then checked out."

Aldrich well summarizes points 5 and 6. He reports that "by 1974 only 12 percent of the population relied on media other than television or newspapers as a principal source of news"; that candidates design campaigns to garner the best coverage; and that lack of time and space in the national media means that "there is little time or space for complex issues. These constraints are especially true of television, where the need for relatively brief and visual stories means that questions of policy receive little attention." Instead, fully 72 percent of the time allotted by network news programs to presidential nominating campaigns is devoted to hoopla—rallies and motorcades. Moreover, "the media are interested in the 'horse race' aspect of the campaign. Many stories focus on who is ahead, who is behind, who is going to win, and who is going to lose, rather than examining how and why the race is as it is." [37] It follows that *in presidential races voters are likely to be ill informed about candidate policy stances and competence.*

(7) It is commonly believed that *momentum has become important in primary campaigns.* In the view of Gary Hart, for example, success feeds on popularity, which feeds on success: "It is cyclical. If you're low in the polls, you can't raise money, and if you can't raise money, you can't sort of bootstrap yourself up in the polls. You have to win primaries. . . . But if you win a primary, you get charismatic very fast and you get money."

Success thus breeds success. As former Rep. Morris Udall has described the primary process,

> It's like a football game, in which you say to the first team that makes a first down with ten yards, "Hereafter your team has a special rule. Your first downs are five yards. And if you make three

of those you get a two-yard first down. And we're going to let your first touchdown count twenty-one points. Now the rest of you bastards play catch-up under the regular rules." [38]

It is widely held that voters have little information to go on in making their choices in the primaries other than the horse-race and hoopla coverage, the sensationalist, skeleton-in-the-closet investigative reporting, and the choices made by voters in preceding primaries. This paucity of information, combined with voters' weak ties to party organizations, is likely to result in weak attachments to candidates; hence a little new information could significantly sway the opinions of many voters. Since most people are seen as getting the bulk of their information from television, the television industry can have a significant input into selecting nominees and ultimately the president.[39]

It has been argued that the media are biased ideologically.[40] If it is true that television does present a consistent, ideologically skewed view of the world, and if Iyengar and Kinder's experimental findings can be generalized to real-world settings, then the media itself may contribute to the selection of ideologically unrepresentative—outlier—candidates, with potentially perverse consequences for American democracy and the common interests of the American people.

(8) *The McGovern-Fraser and later campaign reforms accelerated the changes in presidential campaigns listed above and brought about further changes in the nomination process.* In my discussion of point 4, I noted that presidential nominations have become more contestable than they were during the machine period. The 1968 Democratic convention created the McGovern-Fraser Commission to address questions of procedural fairness and delegate selection. The commission recommended, among other things, that the party publish convention rules and establish uniform times and dates for delegate selection meetings; that state and local parties lower or abolish participation fees (paid by candidates for delegates to the national party convention in order to get on the primary ballot); that state delegations proportionally represent the vote totals won by the various presidential candidates in each state; and that women, young people, and minorities be well represented in these delegations.[41]

Of course the post-1968 reforms do not represent the first time that the parties tinkered with the way they choose candidates for high office. The Progressive reforms at the turn of the century clearly were an effort to change the nomination process, and, as Reiter notes, the Democrats performed a major overhaul of their nominating process in 1936 (also the first year that widespread opinion polling was used in campaigns), when they abolished the two-thirds rule for choosing the party's nominee.[42]

The post-1968 reforms constitute a move by the liberal faction of the Democratic party to wrest control from the New Deal Democrats. They were instituted to help hammer in the last nails on the coffin of the party machine era. The growing predominance of primaries, for example, made presidential nominees, as Austin Ranney states, "less indebted than ever before to congressional and other national, state, and local party leaders and more indebted to their own organizations and contributors and to voters in the primaries." [43] Reiter notes:

> The most dramatic example of this was the ejection of Chicago's Mayor Richard Daley from the 1972 convention because he had not followed the McGovern-Fraser guidelines in composing his delegation. The primaries also seemed to hurt party leaders by enabling candidates to win delegates by ignoring the leaders and appealing directly to the voters.[44]

In 1952, although Estes Kefauver won the most primaries, Adlai Stevenson, the favored candidate of the governors and the Democratic party organization, won the nomination. Likewise on the Republican side, nominee Dwight Eisenhower was outgunned in the primaries by Senator Robert Taft. Since that election, however, only Democrat Hubert Humphrey in 1968 failed to win either a plurality of his party's primaries or a plurality of votes cast in those primaries but still managed to take the party nomination.[45]

Quantitative evidence of the decline of machine influence at conventions is difficult to assemble. One proxy that has been offered is a measure of delegation loyalty to party leaders. Reiter reports that the proportion of state delegations providing a majority of support for the candidate endorsed by their state governor averaged around 80 percent for 1952-1960 and 1968 conventions on the Democratic side and for 1952, 1964, and 1968 on the Republican side. At subsequent contested conventions, the proportion drops to only half of state delegations.[46]

Eliminating the party bosses also meant abolishing the "unfair" rules that limited delegate participation and the unit rule (winner-take-all primaries and caucuses) in favor of more proportional representation. The latter reform had the desired effect: through 1960, state delegations to the convention frequently voted as unanimous blocs on the first ballot (ranging from a high of 85 percent of delegations in 1932 to a low of 46 percent in 1920), whereas since 1968, unanimity on the first ballot has never involved more than 14 percent of state delegations. Reiter also presents two other measures of state-delegation cohesion that roughly confirm this decline in state party unity.[47]

A number of nasty consequences have been ascribed to these reforms. First, they allegedly have made it more difficult for the parties to unify behind their nominees.[48] Since the success of candidate-centered nomination campaigns depends so much on the acquisition of public endorsements and support from various groups associated with the party, such campaigns necessarily lead to the public airing of policy disputes between rival factions in the party. This public process limits the ability of groups to bargain with one another without threatening the electoral value of the party label (by making them publicly commit to specific policy positions).[49] Second, and closely related to this factionalization of the parties, they promoted the participation of issue-motivated activists.[50] As a consequence of increased factionalization and participation of single-issue activists, it is argued, the reforms have contributed to a greater likelihood of an ideologically extreme nominee.[51] At the same time, however, the reforms have been viewed as bringing about a nationalization of parties.[52]

The sum total of these common observations is said to be that parties and party organizations are increasingly meaningless to the American people; that the media—especially television—have assumed many of the salutary democratic functions once carried out by the parties; and that the media have dropped the ball on supplying the information cues and constraints needed for a well-functioning mass democratic polity. I return to these themes in the conclusion.

Stylized Arguments

The mass of arguments made in support of the stylized facts cited above is truly prodigious. Sadly, however, few such studies have laid out the structure of the logical model necessary to account for these facts. In this section, in order to make sense of the stylized claims, I construct a parsimonious model of the stylized presidential literature—a listing of assumptions necessary and sufficient to account for the main assertions—as a precursor to assessing the validity of the common inferences drawn in the literature for American democracy. This exercise necessarily does damage to some of the arguments made by various authors, but I think there is remarkable agreement on each of the basic assumptions. The point of this exercise is to clarify the degree to which the stylizations conflict with one another and to construct testable hypotheses to evaluate the stylizations that are consistent with the model.

There is no "objective" evidence in favor of a number of the above stylizations, only the "informed opinions" of various authors and professionals in the campaign business. Different people have different perceptions; it is difficult to derive hypotheses about presidential campaigns that can be tested solely on such impressionistic data. Instead, we may have to look for more indirect tests to evaluate some of the stylizations; if they follow deductively from other stylizations for which evidence is available, one may reasonably infer that the unobservable stylizations may be true as well. Thus it is vital that the premises of the model be plausible and that the first order implications be robust to technical tweaks to the basic assumptions.

My basic model has two parts: a set of assumptions about information and another set about political organizations in America, which I consider in turn. The goal of this section is to present a sufficient set of conditions to derive the major stylized observations presented in the section above. These assumptions ideally should be basic and simple enough to be virtually uncontroversial.

Television and Information

The advent of TV, it is argued, has brought a new calculus to politics. The usual reasoning implicit in arguments about the effect of television on politics addresses two sides of the issue—demand and supply. I construct a model of the stylized view of each below, starting with the demand side, based on the following assumptions.

(1) *A picture is worth a thousand words.* This figure of speech is so old and well known that it seems almost heretical to question its merits as an assumption. Pictures can affect viewers' perceptions about the world more efficiently than can words. Campaign managers from the 1970s and 1980s certainly think so. For example, Mike Deaver recalls when television reporter Leslie Stahl did a report "about the fact that the visuals that [the Reagan administration] portrayed simply did not add up to what our actions were, the policy. She may have had a point. . . . [But] Everybody sat there and watched this piece that she put together, three and a half minutes, and watched these beautiful visuals of Ronald Reagan. They [viewers] never got the point she was trying to make, because the pictures were so nice."

(2) *TV presents more pictures per minute than does print.* This is uncontestable. It may be true that TV does not present all that many *discrete* pictures, but that is a semantic rather than a substantive objection.

(3) *TV is a public good for consumers.* Assuming that a consumer owns a TV and ignoring the cost of electricity, broadcast TV information is freely available to consumers, unlike newspapers, which generally must be purchased. Assumptions 1, 2, and 3 together imply that the cost to viewers (in terms of time and effort) of acquiring information from TV is less than the cost of acquiring the same amount of information through print, at least for some range of information acquisition.

(4) *The average person's demand for information is low enough to*

be below the threshold, if one exists, where information from print is less costly than information from TV. The point of this assumption is that people, on average, demand little information in making a decision. The assumption is stated rather technically.

I assume that the viewer garners diminishing informational returns from additional TV viewing. The first half hour of TV news consumed, for example, allows the viewer to get a pretty good handle on the news of the day; each additional half hour yields less and less new information. In other words, I assume that the viewer has very little ability to consume selectively from subsequent TV news shows; viewers chose whether or not to watch a news show without knowing its specific content ahead of time. If, as I will argue below, TV shows are geared to the average viewer who demands little information, news shows broadcast during any given time period (a news day) will have a lot of overlap.

At low levels of consumption, TV viewing often substitutes for other mindless leisure activities, so the opportunity cost of watching TV is probably nil. This indicates that for very high levels of demand for information, TV news viewing would begin to intrude seriously on important activities, such as earning a living, sleeping, eating, procreating, or watching Dirty Harry movies.

Consumers can be very selective in their marginal consumption of printed information, however (if an article is repeated from a previous newspaper, the consumer can choose to ignore it). Hence, print-media consumption should not exhibit diminishing returns.

Note that nothing in this model precludes the possibility that some individuals will read newspapers. Those demanding a high level of information will read the newspaper and also watch television news. If television were not available, average information consumption would likely drop as everyone would have to purchase print media. High-level demanders would still read newspapers, whereas many low-level demanders would be priced out of the market altogether.

These four assumptions (pictures are information-dense; TV is picture-dense; TV news is a public good; average demand for information is low) do bring up an important question about what information is. To many of the writers of presidential campaigning literature, possessing information means having encyclopedic knowledge—being able to regurgitate facts, figures, etc. This is not the only way to think about information and is perhaps the least appropriate way to consider how it is used in decision making, as I will discuss later. In this section I concentrate on what level of information consumption one can expect from the average person. In other words, I treat information as a commodity and assume that the level of knowledge possessed by one individual can be directly compared with that level possessed by another.

One last assumption completes the demand side of the information picture:

(5) *People are rational.* This means they are efficient; when given a choice, all else being constant (the amount of information obtained), they will acquire information though what appears to them beforehand as the least costly means.

It follows from the first four assumptions that TV is the least costly means of acquiring all the information that the average person wants prior to making choices, and assumption 5 then implies that *he or she will get all of this information from television.* These assumptions, however, tell us next to nothing about how well informed the average person will be on any particular subject, nor anything about which subjects he or she will be familiar with. I need, then, to make some assumptions about the supply side of the equation. Assume further that

(6) *TV networks seek to maximize profits.*

TV network profits are, of course, generated by advertising dollars. On the assumption that advertisers want to reach as many potential buyers of their products as possible, it follows that, if information can be treated as a commodity, the TV networks will seek to maximize their revenue by maximizing the audience for their news programs. Thus all the networks will target their primary newscasts to the same audience.

The networks presumably know a lot about consumers' viewing habits. They know that average demand is quite low—the below-average demander of information may watch only one half-hour newscast per day. Since it is costless for viewers to change channels, the networks know they are under competitive pressure to deliver as much news as compactly as possible. They will therefore tend to provide brief stories, each built around a single simple theme illustrated by a soundbite or image. As a result, the major networks' newscasts tend to have the same general coverage and format. Any behavioral norms that might arise among reporters to encourage pack journalism—such as are discussed in Sabato's essay—will simply support this tendency.

Supplying more information than the average person wants increases the networks' costs. If the demand for information equals or exceeds the point at which the marginal costs of consuming additional information from television and from print media converge, the increased costs borne by the network will not translate into increased market share, since the consumer can acquire the additional information from print media at a lower cost. Hence, it follows that the nightly news shows will tend to provide short, image-oriented, superficial coverage of many stories. The average person, in my model, wants a survey of the day's events in a quick and dirty format, not in-depth reporting that would require a longer show to cover the same number of stories. It also follows that local newscasts will tend to expand their news coverage to include national stories, since they compete with the national broadcasts for the low-demand end of the market.

The stylization that TV coverage of presidential campaigns is superficial and soundbite-oriented fits quite naturally into this framework. The networks' style of coverage is based quite simply on the character of consumer demand for information and network responsiveness to market demand. I have shown that the first half hour of news programming will conform quite closely to the stylized views of TV news. Since average demand for information is low, there will be little encyclopedic informational content in the news. It follows that the average person won't know much, according to the commonly used definition of information as encyclopedic knowledge. Whether or not the information provided by this half hour of news is sufficient for most voters to make good political decisions is a separate question, which I address below.

Importantly, it also follows that candidates will want to manipulate the pictures people see on TV. Since the average person knows only a few things, any marginal increase in knowledge has the potential to change significantly his or her opinions about the world. Thus how the news is framed and primed is very important to candidates.[53] Witness, for example, Mike Deaver's preoccupation with visuals: "the picture is everything, in my opinion. . . . I happen to think that the picture, the visual, is just as important as the spoken word."

It follows from the assumptions about profit-seeking and demand that the networks will gear their primary newscasts toward the average viewer. We know that voters are not drawn randomly from the pool of viewers; voters tend to be wealthier, better educated, older, and are more likely to be white than a randomly drawn TV viewer. The interests of the average voter and the average viewer are therefore unlikely to coincide. It follows that the level of information provided by primary TV newscasts will be lower than the average voter would prefer. The substantive content of those newscasts is also likely to be skewed away from the interests of the average voter, so that voters will be getting biased political information. Relatively more coverage will probably be devoted to candidates and issues of interest to the average viewer, relatively less to candidates and issues that would be of greatest interest to viewers with more extreme views and interests.[54]

Nomination campaigns are necessarily concerned with that portion of the electorate likely to vote in each party's respective primary elections. It is hardly controversial to assume that there is a difference between the average Democratic and Republican primary voter, and that both differ from the average person in the electorate as a whole. The average Democratic (or Republican) primary voter will know more about candidates whose positions are closer to the interests of the average television viewer rather than about candidates whose positions better satisfy the voter's own interests. It also follows trivially that, relative to the issues covered by the campaigns and the requirements of various partisans, TV coverage will be biased.[55]

Under certain circumstances, then, networks may influence the choice of the party's nominees and the president. If voters are risk averse, they will be more likely to vote for a candidate they know and dislike somewhat rather than a candidate they do not know at all. Voters make inferences from the information signals given to them by television—that some candidates' positions are relatively close to the interests of the average television viewer. Voters know only that the positions of other candidates are not close to the average viewer's interests. Such candidates could, in fact, be right- or left-wing extremists or somewhat more moderate, but more precise information is unavailable, and the risk-averse voter will prefer the small evil who is known to the potentially large evil not known.

I have been able to derive most of the things attributable to the advent of TV from six basic assumptions (plus a few additional technical assumptions): (1) pictures are information-dense relative to printed and spoken words; (2) television is picture-dense relative to other media; (3) television provides cheap information; (4) people don't need full information in order to make choices in the world; (5) people are efficient in their information-consumption choices; and (6) competitors in the TV industry want to maximize the relative size of their audiences in order to make money. I now build upon this foundation, adding assumptions about party decline.

Primaries, Candidates, and the Decline of Parties

The decline of parties in twentieth-century America is an article of faith for most political observers. In an earlier time, political machines ruled the roost; today they do not. The heart of our culture's definition of political party is the notion of machine politics, characterized by a tightly disciplined, hierarchical organization headed by one person or a small cabal of leaders, in the mold of Boss Tweed of Tammany Hall, Richard Daley (the elder) of Chicago, Don Corleone of Mario Puzo's New York, and the Galactic Emperor of George Lucas's imagination. Machines did three things: they restricted entry for candidates seeking office and regulated turnout and voter choice at the polls.

The decline of political machines is often traced to changes in electoral laws, which affected their ability to determine electoral results, and municipal reforms, which reduced their incentives to control local governments. Most machines had local roots, controlling a city or county government and the jobs and contracts that such control could provide to its foot soldiers. The Progressive municipal reform movement of the early twentieth century transformed local governments, often increasing the accountability of elected officials to the voters. Career, professional bureaucracies were established in many larger cities, which allowed government workers in some cases to organize themselves on a separate basis from the political machines (e.g., to unionize).

Even more important, the secret Australian ballot made it almost impossible for the machine to ensure that the voters it brought to the polls would faithfully cast their ballots for machine-endorsed candidates. The use of government-printed ballots and formal procedures for getting one's name placed on the ballot also changed the locus of electoral competition, allowing competing interests to challenge the established party machines. The net result of these developments was the withering away of the hegemonic position of party machines in American elections.

Since political machines, for the most part, no longer exist in the United States, it follows that they no longer control the presidential nomination process. More significantly, political machines no longer impose great barriers to entry for competing candidates in the nomination process, nor do they determine turnout or voter choices at the polls. Rival organizations with well-known reputations, such as unions and interest groups, now provide candidates with endorsements that imply electoral support from voters, enabling nonmachine candidates to run for (and have a chance to win) the presidency.

Further, since general election ballots present voters with a choice of candidates for many different offices, since no nonmachine central party authority can force candidates to help pay for getting out the vote for the ticket, and since no organization can guarantee how any particular voter will cast his or her (secret) ballot, it follows that no organization will have an incentive to spend very much on get-out-the-vote efforts. As a result, turnout has declined from its peak during the machine era. Note that this effect has nothing at all to do with television.

Once a voter gets to the polls, how will he or she vote? Without a party machine to make voters support a whole ticket, other voters can't be depended on to ignore a candidate's individual characteristics and vote solely on the basis of party or party faction. Therefore, the voter's best bet is to do the same: cast a vote for each office independent of the choices made for other offices. More voters, on average, will thus split their tickets at the polls, especially if they want different things from different offices. And finally, from the perspective of individual voters, it follows that survey respondents, when asked to summarize their political views on different offices and institutions with a single indicator, will be less likely to identify themselves as strong partisans. A smaller share of voters will take this position than would have so identified themselves if they had no incentive to pursue different voting strategies for different offices.

The main observation that remains to be explained is the rise of candidate-centered campaigns. By candidate-centered, I mean that for each candidate for a particular office there exists an organization dedicated solely to getting him or her elected; and that the questions and issues of the election revolve around the competences of the candidates and their policy positions. One prominent explanation for the rise of candidate-centered elections is the introduction of primaries.[56] However, primaries by themselves do not produce candidate-centered general elections. They are clearly not necessary, given that many local elections—e.g., school board, county superintendent, municipal judge, and city council—for which primaries are often not required are often candidate-centered and not party-centered. Moreover, numerous party machines during the twentieth century successfully directed votes and ran elections, including primaries. The most prominent example of a machine thriving despite primary elections is probably the Daley Chicago machine of the 1950s, 1960s, and 1970s.

There are three broad conditions sufficient to yield candidate-centered presidential nomination campaigns. The first is that candidates are able, at low cost, to gain widespread recognition in order to be considered by a majority of convention delegates. All states have more or less restrictive requirements for having one's name placed on primary ballots, so entry is not free, but once that fee is paid, the candidate is assured of being known to every primary voter.

Second, candidates must be able to create a coalition of voters who choose the candidate based on his or her personal reputation rather than solely on the reputation of any group to which the candidate might claim allegiance, such as a party. Third, this coalition must be potentially decisive. I discuss the latter two conditions in turn.

Aldrich, in an essay later in this volume, addresses the role of personal-vote coalitions. He points out that elections can become candidate-centered only if a technology exists that allows candidates to create a personal brand name. He observes that for presidential elections this technology did not exist until TV became a mature medium in the last two decades. But what does it take to create a brand name, or at least one large enough to encompass a potentially decisive coalition of voters? Perhaps we should first ask what a brand name is. It is a cue or signal that conveys information about the quality (and other relevant characteristics) of a product—in this case, a candidate. Perhaps the best examples of brand names are McDonald's and Coca-Cola. At the mere utterance of the name "McDonald's," people recall its products and service with remarkable accuracy and detail. The strength of this brand name, however, is not the result of TV advertising, as Aldrich's argument would imply, but rather of experience. Indeed, both the McDonald's and Coca-Cola brand names existed prior to the maturation of TV as an advertising medium.

Neither company's TV commercials today typically provide any information about products or service. Instead, their ads feature celebrity singers, catchy slogans about "Food, Folks and Fun," and teaching the world to sing. Likewise, for a well-known candidate (such as an incumbent), there is little need to provide such information, since we all know it already. Conversely, for little-known candidates (and new products), advertising alone is not sufficient to create a brand name.

In the world of politics and politicians, I would argue that personal brand names have been around for a long time. For example, nearly every voter at the time knew something about George Washington, Thomas Jefferson, and Andrew Jackson. Indeed, their brand names were so strong that they still operate today. Most Americans probably could not begin to name specific policies advocated by Jefferson, but they could say something about what he stood for; moreover, they might know something about his personal life, in much the same way that Americans have some knowledge of politicians' personal lives today.

Many companies, with products ranging from automobiles to amusement parks, have established brand names for their products. These brand names—with a reputation among consumers for the quality and the characteristics of the products—are quite valuable and fiercely protected. The difference in price between brand-name drugs and their generic counterparts reflects largely the value of a brand name to the product's manufacturer. Knowing the reputation of brand-name products helps reduce the amount of encyclopedic information consumers need in order to choose between products. Indeed, most people use little more than a brand name to help them choose among big-ticket purchases such as cars and home appliances.

What makes brand names such as McDonald's so valuable is not the TV ads but rather the fact that nearly every one of us has been to a McDonald's and we have all shared the same experience. With cars, televisions, bedding, and toasters, brand-name reputations are created through our personal experiences with various products—or through testimonials from people who claim firsthand experience with a product. We might believe, for instance, that consumers can accurately infer the relative quality of competing products based on only two bits of information: prices and sales data.

There is, however, one critical difference between consumer-product brand names and candidate brand-name formation. Providers of consumer goods can compete on the basis of price, whereas candidates cannot. Voters presumably don't care who wins office, only what effect different officeholders will have on policy outcomes. Since price is assumed to be fixed, if retaining the incumbent is one of the choices, it follows that voters will not choose another candidate who promises identical policy outcomes, all else being equal. Challengers must advertise that they are different from the incumbent on one or more policy dimensions. Voters will be able to infer from a challenger's level of expenditure that he or she offers outcomes that will differ from the status quo.[57]

Candidates develop reputations for championing certain policies or segments of society. Hence, John Petrocik tells us, not all advertising claims are credible.[58] David Duke can no more erase his past affiliation with white supremacist organizations than Jerry Brown can escape his offbeat reputation, or Walter Mondale his image as an old-line, "special interest" Democrat. If a claim is contradicted by experience, people tend to discount the new claim—or, perhaps worse for the candidate, devalue the standing reputation. Once created, a reputation is both a valuable campaign resource and a constraint on a candidate's behavior. Television advertising can be used to reinforce the brand name once acquired, but it cannot be used to create one. At most, it can reduce differing perceptions of quality among consumers by exposing large numbers of them to an identical endorsement. The much vaunted incumbency advantage, of course, underscores this point: incumbents rely on their personal brand names to deflect competition and win reelection (although, interestingly, incumbents almost always outspend challengers in congressional races).[59] Given my assumptions about information transfer, the easiest way for endorsers to convey their endorsements to a large decentralized audience is through television[60] (but, again, TV advertising is neither necessary nor sufficient to create a brand name).

The chain of endorsements through which a reputation is established may be quite long. Typically, however, it extends from the candidate to some set of individuals with very large (well-known) reputations (such as interest groups, ex-presidents, Ralph Nader, or a political machine organization) to individual voters. A personal-vote coalition is simply a set of individual voters who independently choose to support a particular candidate based on their personal experience with that person. This coalition differs from other kinds of support coalitions in that it implies no strategic interaction among its members. In contrast, a machine-based coalition is a cartel of voters—a set of individuals who explicitly seek to coordinate their vote choices with one another. A party or issue voter, on the other hand, chooses a candidate based on the endorsement of a person or group whose reputation the voter already knows, such as a political party or interest group.

Returning to the explanation of candidate-centered general elections (and proceeding from the six assumptions about information given in the previous section), in order to create a deductively valid explanation for this purported fact we need to assume that:

(7) *Candidates have personal brand names at the beginning of the general election.* My purpose here is not to write a book about brand names. Since they are a central part of most people's discussions of campaigns, of who wins and who loses, I will simply make the assumption and move on.

The third condition for candidate-centered campaigns basically states that there must be some gain to be made for a candidate to center the campaign on his or her own brand name as opposed to the party's brand name (or as opposed to the wind and the stars). The gain, of course, is in votes. Thus we assume:

(8) *There are voters whose choices are not predetermined; and this set of voters is a decisive set.* By this I mean simply that if all members of the set vote the same, their choice determines the outcome of the election. Hence, for the set of all voters whose choices are not "hard-wired" to be decisive, the predetermined votes cannot give a majority of votes to a single candidate, nor an insurmountably large plurality if there are more than two candidates.

Since party identification is said to be the principal source of vote predetermination, making partisan appeals is unlikely to sway the critical, nonpartisan swing voters. Thus, given the above model of electoral competition, it follows that general election campaigns will center on candidates rather than parties. It also follows that primary campaigns will focus on candidates' attributes and interest-group endorsements; since all candidates in a primary are competing for the same party nomination, there are no predetermined votes.

Developing, maintaining, and reinforcing personal brand names (i.e., keeping to a minimum variations in people's experience with the candidate) are costly. This is clearly the case in the business world. McDonald's expends considerable effort and resources in conducting oversight of its franchises—training seminars, periodic inspections—and in supplying the characteristic McDonald's containers. Another example of the cost of maintaining the value of a brand name can be seen in the U.S. automobile industry, where manufacturers periodically issue voluntary recalls for an entire model year to repair some systematic flaw. As American manufacturers have slowly learned in the past twenty years of competition with Japanese firms, consumers make inferences about product lines based on very little data, such as a bad experience with a single product. In the 1970s and 1980s thousands of American car buyers turned their backs on American producers in part because of their belief that Japanese cars had superior quality control.

It follows from the high costs of establishing and maintaining brand names that candidate-centered primary and general election campaigns will be long and costly. This is because candidates must establish themselves either with a set of endorsers whose reputations are very well known or with voters directly at the grass-roots level. Endorsers have their own reputations to protect, so they will not grant endorsements freely. A candidate has to find some way to make a precommitment to serving an endorser's interests. Reaching voters directly, on the other hand, is a logistical problem that requires both time and money.

During the primaries, if there is no candidate with a national brand name, then the candidate who can first acquire one will have a big advantage. Winning a state primary election sends a signal to voters in other states that this candidate, for whatever reason, was attractive to a large number of voters. If we stipulate that media coverage of elections includes opinion polling and exit polling, it follows that the media will treat primary elections as horse races.

Note that this is true with or without television! Furthermore, for party voters (some 60 percent of the electorate, according to usual measures), candidates who can win their party's state primary elections should be attractive prospects to bear the party banner in the general election. Thus winning—or at least doing well enough to help establish a brand name—is important. Momentum is important, but not in the simplistic sense that it creates unbeatable juggernauts. Rather, people take cues from others whose reputations are well established: either individuals whose actions we have observed or representative samples of demographic groups, on whom we also have observations.

I agree with the assessment that momentum, observed as an

apparent bandwagon effect, can be important, but only in limited ways. As Geer puts it,

> The crucial component of this bandwagon scenario is that most voters have little information about the candidates. When voters have little or no information, any new information will create images about the candidates. Given the positive coverage that follows a winner, this initial impression of the candidate is likely to be favorable. . . . Of course . . . if a relatively unknown candidate begins to receive bad press, voters may start to respond less positively to that contender.[61]

Bandwagons of this kind, therefore, are unlikely to occur in races where the competing candidates are generally well known. When this is the case, voters will generally have more stable preferences about the candidates, making large shifts in loyalties less likely, holding constant the field of candidates. Of course the news media's coverage will have some influence in races featuring well-known candidates, since some voters will have weak preferences or none at all. My point, however, is that there are not enough uncommitted or weakly committed voters for marginal bits of information to generate large shifts in candidate support.

Lastly, the rise of political consultants is no mystery, and an explanation of their ascendancy need not say anything about the decline of parties or the sinister nature of Madison Avenue. Rather, changes in technology lead rational candidates to hire experts in using the technology. Indeed, this was true with the advent of mass newspapers, which led candidates to hire speech writers and press secretaries. Likewise, the advent of TV led candidates to hire ad men.

The Reforms

Still to be discussed are what effects, if any, the post-1968 reforms in the nomination process have had on presidential selections, elections, and policy outcomes. The major reforms, as noted above, changed the procedure for selecting delegates to nominating conventions to provide better demographic representation of state populations and to represent more accurately the voting strength of presidential contenders in states holding primaries. A number of states also switched from caucus or convention formats to primary elections for committing national convention delegates to presidential candidates.

The standard interpretations generally argue that the reforms led to more ideologically extremist participation in state and national party conventions, more influence being given to narrow, well-organized party factions or interest groups, and a general lessening of the cohesiveness of the major parties. Unfortunately there is very little evidence that (a) the reforms caused changes in the ideological composition of the party conventions, or (b) that changes in the composition of the convention caused or reflected a change in the kinds of candidates being nominated.

My approach to interpreting the effects of these reforms is to extend and modify the basic model presented above. In general, groups—demographic, ideological, or particularistic "interest" groups—are the focus of most studies of reforms of the nomination process. Therefore, I need to make some assumptions about how such groups are involved in American politics.

By a group I mean an organization—a collection of individuals bonded together by some institutional structure to pursue a common goal, be it profit, mutual entertainment, or pressuring the government for the redress of grievances. My definition does not include such collectivities as social movements or any other collectivity united only by a common demographic, economic, or behavioral trait—i.e., blacks, whites, men, women, Jews, Christians, or small businesses. My definition does, however, include demographically homogeneous organizations such as the NAACP and the National Organization for Women, as well as gardening clubs, book clubs, churches, and so forth.

The organizations that provide structure for collective efforts and goals are often well placed to acquire information relevant to presidential candidate selection and to provide this information, in the form of cues, to their members. Indeed, if the group's reputation in the political arena is well known, its cues may affect the behavior of a much wider circle of individuals.[62] I have already assumed that information is costly to acquire and that as a result people will be "ill informed" in an encyclopedic sense, all else being equal. The presence of groups may change this calculation. At a minimum, groups mean that some individuals will substitute cues for encyclopedic knowledge—what I described above as the general understanding of "information"—and thus will be better able to make "informed" decisions with respect to many choices, such as which candidate to vote for in a primary.[63]

Under certain circumstances, when the issues in an election affect the collective interests of a group, it will be motivated to provide not only an information cue to its members (and others with access to the group's cues) but also selective incentives to turn out and vote.[64] Such groups, in other words, will try to vote as a bloc or cartel, just as political machines did in the past. These selective incentives and information cues are especially important in primary elections. In a primary people often have very little information about the candidates, who tend to be not very well known (especially for the nonincumbent party); the party label provides no information since all of the candidates share the same one. If we assume that voters are risk averse and less likely to vote if they lack sufficient information to make good inferences about the effect on policy of a given candidate's victory, it follows that members of the set of policy-interested groups are more likely to vote in primaries than voters who are nonmembers.

Thus primary voters tend to be "interesteds"—people with affiliations to groups with particular policy axes to grind. There are two kinds of organized, politically active groups: those that favor the status quo policy, all else being constant, and those that reject it, who are therefore willing to bear some cost to get the policy changed. Assume that this is costly. Since I have assumed that voters are risk averse, groups whose interests are already being served by the government have no need to be active to maintain the status quo and groups whose preferred policy differs only slightly from current policy probably can't justify the expense to change policy. Groups seeking sharp policy changes, on the other hand, are more likely to work actively to change the status quo.

Hence, there are two major categories of primary voters: either a supporter of the status quo or an advocate of radical change. It follows that, if voters' preferences on the relevant policy dimension are uniformly distributed, voters in the nonincumbent party's primaries will tend to be extremists (relative to the status quo policy), and extremist candidates will tend to win primaries and, therefore, convention delegates. These delegates, I assume, will foresee that general election voters are risk averse and would not replace an incumbent with a challenger who promises more of the same. Hence, the Democrats in recent years have tended to nominate very liberal candidates for president.

Primary voters in the incumbent party, on the other hand, tend to be supporters of the status quo policy. Given a choice between an incumbent (who promises to maintain the status quo) and an intraparty challenger who proposes nothing more than better management of the status quo policies, voters would have to believe the incumbent's mismanagement was tremendous for them to throw that person out of office. Hence, incumbents almost always win renomination, except when they get caught in scandals.

If during the nonincumbent party's primary process enough delegates are committed to a single candidate to win nomination on the first ballot, it follows, as night follows day, that the convention will tend to nominate extremist candidates, in the sense that they will differ sharply from the incumbent on at least one policy dimension. If voters fail to coordinate in this way, there is no obvious stable solution to this process. Groups may logroll to nominate a candidate who is extreme on multiple dimensions or they may compromise to nominate a candidate who differs relatively less from the status quo on those dimensions. Happily, since the installation of the reforms in 1972, no nominating convention has failed to select a candidate on the first ballot.

A Matter of Facts

I have surveyed the most common observations about modern presidential elections, offering an explanation for them based on the most common premises in the scholarly literature. Having character-ized people's beliefs and analyzed their origins, I now question the reasons why people believe these things: i.e., what evidence can we find that the things people seem to believe are, in fact, true?

This question, of course, brings up another: what constitutes evidence? My preference is for reproducible outcomes from repeated experiments. Failing to find many laboratory experiments regarding peoples' beliefs about presidential elections, I prefer quasi-experiments, that is, experiments in nature where only one (or very few) of the factors affecting the studied phenomena is changed, enabling us to observe directly the effect of such changes on the studied phenomena. Price changes for products and commodities provide quasi-experiments, for example, that allow us to test economic theory (does demand for a product rise or fall as its price increases?).

It is possible to conceive of some quasi-experiments on modern presidential election campaigns. I outlined three suitable hypotheses above: that the decline of party machines caused specific changes in presidential election campaigns; that the advent of television caused other changes; and that the McGovern-Fraser and subsequent reforms caused additional well-specified changes. I might, of course, have trouble specifying when a change took place and when its consequences for presidential elections were manifest—when exactly was the advent of television, and when did we expect to see presidential elections change as a result?—and the authors of the literature surveyed above give me little (or no) guidance. Nonetheless, I could, in principle, construct some quasi-experiments.

Frequently, however, political observers do not collect anything like experimental or quasi-experimental evidence. All too often, no objective evidence whatsoever is given for the purported phenomena: no evidence that I could easily reproduce on my own; no widely accepted, contemporaneous measures of the phenomena; and no databases from which I might draw evidence. Rather, the evidence given is subjective and personal, constituting an opinion or judgment.

The problem with such evidence, however, is that on the one

hand, I hate to throw it out, particularly when many smart people have come to the same opinion based on their personal observations. On the other hand, as we all know, opinions vary. There are often at least as many opinions concerning each observation as there are people who observed the phenomena. To cite a specific example, even if all observers agree that journalists have shown less restraint in the types of stories about politicians they pursue and file, there is no agreement (and indeed no discussion) on how much less restraint they now show (5 percent? 10 percent? A little? A lot?). For these and other reasons, I will not give much credence to measures based on observers' opinions.

For the most part, the stylized facts enumerated above are backed by no objective evidence. Rather, they are observations made by campaign insiders and analysts, agreed to by all concerned and given credence in print or by repeated use on TV. I take these observations seriously. This does not mean, however, that I regard them as true facts but rather as things that need to be explained: why do these people believe things that, when held up to an objective standard, either cannot be confirmed or can be positively rejected?

Just the Facts, Ma'am

My first demand is that the stylized facts surveyed above *be* facts. I will take them in the order presented earlier.

(1) *The power of party machines to control votes, and therefore to control the nomination, has disappeared.* Outside of certain variants of Marxist conspiratorial theory, everyone believes that the state and city machines of the nineteenth and early twentieth centuries no longer exist and have not been replaced by new machines. What evidence could one present to show that something does not exist?

While I accept this "fact" as true, it is difficult to say when it happened; and since it is purported to be an intervening variable that generates further changes in presidential elections, it is important to be as precise as possible as to its genesis. The decline of party machines is most often associated with the Progressive reforms at the turn of the century.[65] I have argued that political machines were, in large part, cartels of voters. Cartels are, of course, difficult to maintain in the face of contrary individual incentives. The ballot reforms of the 1880s and 1890s and the later Progressive reforms made those cartels harder to maintain by reducing the observability of voting behavior and reducing the availability of patronage jobs in many cities. We have some indirect evidence supporting the claim about the decline of state and local machines from voter turnout figures, which I discuss below.

We have no clear-cut answer, however, as to when machines declined in importance for presidential nominations. Most of the percentage decline in turnout for presidential elections occurred in the first two decades of this century. Paul David and his colleagues conclude, with regard to the Democratic and Republican nominating processes in 1952, that " 'very few of the state political conventions of 1952 were within the grip of a recognized state political boss' and that 'the old-time local political bosses seem to be dying off without being replaced.' " [66] It is often argued that the bosses held their grip on presidential nominations by controlling the selection of delegates to the national party conventions. However, if we assume a world in which all bosses control blocs of delegates, but only some control blocs of voters as well, which bosses would we expect to wield the most clout in the smoke-filled rooms? Bosses without electoral clout quickly cease to be bosses.

(2) *Successful candidates have their own large and expensive campaign organizations.* The first question, of course, is what is meant by large and expensive? In comparative terms, presidential campaigns are not all that expensive. In 1984 Mondale and Reagan spent a combined $80.8 million on the general election campaign, or about 87¢ per voter. Prenomination expenditures by all major party candidates in 1984 totaled another $132.7 million, or $1.43 per voter. In comparison, House candidates in 1984 spent $177.6 million, or $1.92 per voter.[67] Americans are pikers compared to the Japanese, however. Estimates of campaign spending by LDP candidates alone reach as high as $50 to $120 per voter.[68] In comparison to Japan, U.S. presidential campaigns are cheap.

Whatever we think about the size issue, candidates have at least had their own organizations since the advent of primaries. But so did Andrew Jackson, around whom Martin Van Buren built the Democratic party in the 1820s. Similarly in 1848 Van Buren built a Free Soil party organization around his own candidacy, winning 10 percent of the popular vote; former Whig president Millard Fillmore carried 21.5 percent of the vote as a Know-Nothing in 1856; Teddy Roosevelt finished second with 27.4 percent of the vote in 1912 under a Bull Moose party label created from thin air. More recently, goes the argument, presidential campaign organizations have become complements rather than alternatives to party organizations. John F. Kennedy, for example, was said to have run a highly structured (albeit small) campaign organization that was not linked to the Democratic party organization[69]—at the very beginning of the television era and before primary victories alone could assure a candidate the nomination.

Within this category it is also alleged that consultants have replaced party leaders in key campaign roles. People who have been involved in campaigns, such as Gary Hart, Susan Estrich, and others participating in the symposium on "Campaigning for the Presidency" held at the University of California at San Diego in 1991; and firms such as Joseph Napolitan Associates, Cambridge Survey Research, Market Opinion Research, Bailey/Deardourff and Associates, and Decision Making Information—to name just a few—are the best evidence for this.[70] Again the question is, When? Keech and Matthews described the 1940 campaign of Wendell Willkie, for example, as "a classic, highly professional public relations campaign," led by "talented and highly experienced experts in the communications industry and public relations"—long before the advent of the TV era.[71]

It is hard to measure meaningfully whether presidential elections have become candidate-centered. Wattenberg performed a content analysis of election coverage in a limited selection of newspapers and magazines (not television) and found that "throughout the whole 1952-1980 period, mentions of candidates outnumbered those of parties, but . . . the ratio increased from about two to one in the 1950s to roughly five to one by 1980." [72] This evidence is consistent with the hypothesis that presidential elections have become increasingly candidate-centered. I find substantial support for this hypothesis in the literature.

(3) *Voter participation and party identification have declined.* A decline in turnout is undeniable, but, contrary to much opinion in the press, the drop came *before* 1912: turnout peaked in the period 1856 to 1900 when voter participation in elections averaged 78.5 percent (standard deviation 3.1 percent); it declined to slightly more than 65 percent in the 1904 and 1908 presidential elections and to 59 percent in 1912. Since Woodrow Wilson's first election through 1988, however, turnout for presidential elections has averaged 58 percent of the estimated eligible electorate (standard deviation less than 5 percent), with the lows obtaining in the 1920s.[73] Turnout has clearly declined from the highs of the late nineteenth century, but the bulk of the decline actually occurred between 1896 and 1912.

As I mentioned above, the decline of party identification is a somewhat controversial issue. If we rearrange Wattenberg's subcategories to group "leaners" together and place "independents" with those who reported no preference, the data show that the National Election Studies' proportion of nonaligned voting-age respondents ranged from 10 to 15 percent for the 1968-1980 period. Leaners ranged from 18 to 21 percent of respondents. Making inferences from these data to the voting-age population as a whole, of course, requires that one compare these ranges of variation with the margin of error implicit in the polls, typically between 1 and 2.5 percent.[74] Thus it is not at all clear that these data show any significant trend toward decreased partisanship. As Jacobson points out,

The degree to which partisanship has declined is a matter of some controversy. If partisanship is measured as the proportion of citizens who claim allegiance to one of the major parties, excluding apoliticals and those who call themselves independents even though they admit, on subsequent questioning, to lean toward a party, data from the American National Election Studies show a decline in party identifiers from 74.4 percent of citizens in 1952 to 62.7 percent in 1988. If leaners are included among the partisans . . . the decline is considerably more modest, from 91.1 percent in 1952 to 87.8 percent in 1988.[75]

Thus if we are to take stock in survey data, it becomes a matter of taste and objectives whether one emphasizes the relatively large decline in strong identifiers or the minuscule drop in the more inclusive measure of identification.

Even if party identification has declined, it is highly questionable whether this means that parties no longer serve a purpose. As to the decline of party identification, the percentage of the electorate who are party-line voters (whose choice for president aligns with their party identification)[76] has, in fact, increased since 1952: 77 percent of the electorate were party-line voters in presidential elections in 1952; this increased to 79 percent in 1960 and 1964 and decreased to 69 and 67 percent in 1968 and 1972, 73 percent in 1976 and 68 percent in 1980. The percentage of party-line voters in the 1984 and 1988 presidential elections, however, increased dramatically to 79 and 81, respectively. And while there has been a slight decline in the percentage of party-line voters in Senate and House races, from the low eighties in the 1950s to the low to mid-seventies in the 1980s, party identification is still the single best predictor of the vote in these races as well.

Moreover, even if we come up with some interpretation of the evidence, or new twists in the data, that purports to show a decline in party identification, does this necessarily imply that parties have declined? If we think of parties as machines, with voters following the marching orders of machine bosses, then a decline in discipline—a drop in PID—would, in fact, correspond to a decline in machine (party) strength. Thus a drop in PID implies a corresponding decline in the ability of party machines to organize elections.

But thinking of parties as nineteenth-century machines is not a particularly useful way of understanding the role they play in modern politics. Though the analogy is not perfect, the most useful way of thinking about modern parties is as franchise organizations. Much like the McDonald's brand name, party labels convey information—cues about the policy positions of candidates to voters.

(4) *Nominations are more contestable than they were in the era of*

strong parties and bosses; outsiders and nonmachine candidates now have a chance to win the nomination. The fact that most candidates drop out of the nomination race well before the convention suggests that observations of increased contestability are illusory. Indeed, one might argue (although I will not do so), based on the modern trend toward conventions as coronation ceremonies rather than nominating contests, that nominations have become *less* competitive in recent years.[77] The proportion of first-ballot votes won by minor candidates (those receiving less than 10 percent) has fallen from the 1896-1932 mean of 28.9 to 15.1 percent in 1948-1960, to 7.4 in 1968-1984.[78] This makes it sound as though the prereform nomination process was eminently fluid and contestable. And Reiter further notes:

> There are problems in testing the hypothesis that there are more candidates entering the race today than before. . . . Many a candidate in the past waited in the wings, hoping to be regarded as a latter-day Cincinnatus and plucked from his plough. Conversely, there are literally hundreds of announced candidates, especially now that the Federal Election Commission requires a statement of candidacy before one can collect contributions.[79]

There is good reason to believe that party machines and voter turnout have declined since the Progressive reforms and that campaigns have become more candidate-centered. There is, however, little support one way or another for any contention about changes in party identification or the contestability of nominations. I now evaluate the evidence on the alleged effects of television.

(5) *Voters get most of their information from television.* Basically it has been true since 1963 that more people get most of their *news* from TV. The percentage of people citing this source in response to a survey question has been between 64 and 67 since 1972, while the percentage who cite newspapers as the main source of news has declined from 57 in 1959 to 42 in 1988. (Note that people could answer both TV and newspapers, so that the percentages can, and do, add up to more than 100.)[80] At the same time, people have found TV news to be more credible; in 1988, 49 percent of the people surveyed were "most inclined to believe" TV reporting, while only 26 percent answered the same question with a preference for newspapers. These survey data, however, do not address the question of whether people get most of their information from television, where information refers to both encyclopedic knowledge and endorsements, impressions and opinions (cues and signals).

A 1984 survey found that 58 percent of the respondents read more than two newspaper articles about the presidential campaign, while 23 percent read none at all. On the other hand, only slightly more—62 percent—watched more than two TV programs about the campaign. More than half of these same individuals admitted to paying very little or no attention to newspaper articles about the campaign, while only 31 percent admitted to paying no attention to TV programs.[81] These data indicate that most voters exert little effort to collect information from the media about presidential campaigns.

(6) *The media and especially TV ignore substance in favor of horse-race stories and muckraking.* Brady and Johnston, in an analysis of media coverage of presidential candidates, judged some 31 percent of the volume of candidate coverage to be "serious," i.e., having to do with "electability, experience, leadership, personal qualities, and policy positions." Approximately half of that serious coverage dealt with candidates' policy positions. On the other hand, some 22 percent of the volume was judged to be "less informative," such as stories about candidates' campaign appearances.[82]

Stanley and Niemi report the results of another analysis of TV news coverage in the 1988 presidential election: 50 percent of the coverage during the primaries was related to horse-race stories, while the other half of the 1,064 stories aired during this time were about campaign (focusing on the candidates) and policy issues. Coverage of the general election, however, was much more heavily skewed toward campaign issues (46 percent of the 1,237 stories aired) and policy issues (36 percent), with only 19 percent of the stories about the horse race.[83]

Does it follow that voters will be ill informed? The data suggest not only that television and print news provide far more nonfrivolous information than many observers would have us believe but also that information acquisition is a much broader process than reading newspapers and watching television. As Popkin notes, people obtain information about the economy and government policies through their everyday activities.[84]

People have simply thought the wrong way about information. It is not merely encyclopedic knowledge. It is also clues, signals, and shortcuts that enable people to make accurate inferences on which to base their decisions. For example, do people need to have encyclopedic knowledge about basketball shoes to choose between Nike and Keds? Or is it enough to know that professional (and perhaps collegiate) basketball players wear Nikes and not Keds?

We acquire encyclopedic knowledge about very few of the decisions we make. We learn very little about the cars, televisions, or stereos that we purchase, relying instead on brand names and endorsements from reputable sources, such as consumer magazines. Why should voting decisions be any different? Indeed, why should we exert as much effort in voting as we do in choosing basketball shoes, given that the probability is close to zero that any individual's vote will tip the election? What voters want is a cue or signal that allows them to make accurate inferences about expected performance. Party labels are one such cue. As Anthony Downs argued, if party labels are linked to policies in a consistent way,

> some rational men [may] habitually vote for the same party in every election. In several preceding elections, they carefully informed themselves about all the competing parties, and all the issues of the moment; yet they always came to the same decision about how to vote. Therefore they have resolved to repeat this decision automatically without becoming well-informed, unless some catastrophe makes them realize it no longer expresses their best interests. . . . [T]his . . . saves resources, since it keeps voters from investing in information which would not alter their behavior.[85]

Personal brand names provide another cue, which function in the same way as the party label. For most people these two cues will be sufficient. A further segment of the electorate will be swayed by individual or interest-group endorsements, which the candidates themselves will happily advertise. The end result is that most people are probably able to make quite accurate inferences about candidates and their preferred policies on the basis of only a very few signals. This allows them to make reasonably good decisions (which under certain circumstances approximate full information decisions) without having encyclopedic knowledge.[86] Gary Jacobson, for example, argues forcefully that voters know, at least generally, the types of policies advocated by each party, and that they vote for Republican or Democratic candidates for different offices, depending on their perception of the best type of candidate for each.[87] Individuals, when asked which party cares more about (or is best able to deal with) given groups or issues, seem quite able to differentiate between the two parties on matters of substance.[88] That most people have little or no encyclopedic knowledge is no surprise. The surprise is that analysts ever expected them to have it.

While the presidential campaign literature generally contends that print media and TV news are different in content, there is no evidence at all to support such a claim. Trivially, of course, there is a content difference, since print does not offer moving pictures to supplement text, but how would one determine that the informational value is different? How people actually use the informational content of TV as opposed to print has not been studied. However, if, as I have argued, Popkin's and Lupia's approaches are correct—that what matters in news are the inferences that signals make possible rather than some pedantic accounting of the number of facts a story contains—then the common criticism that the text of a TV newscast wouldn't fill the space above the fold of a newspaper's front page is irrelevant.

(7) *Momentum matters.* Evidence has been presented by Geer, for example, that purports to show the importance of momentum. For four "Unknown Candidates"—Bush in 1980, Carter in 1976, McGovern in 1972, and Hart in 1984—support from their party's voters increased rapidly upon becoming "known," with jumps ranging from 10 percentage points for McGovern to 27 percentage points for Hart (though only for Carter did the level of support increase above 50 percent); while for nine "Known Candidates"—Humphrey in 1972, Muskie in 1972, Ford in 1976, Reagan in 1976 and 1980, Carter in 1980, Kennedy in 1980, Jackson in 1984, and Mondale in 1984—support among their party's voters showed no upward trend, only slight month-to-month variations over the course of the campaign.[89]

But what does this tell us? First, some perspective is needed. For the most part, the support levels obtained by the "Unknown Candidates" were near zero before they became "Known." After they became known, their support levels (except for Carter) were still largely below those enjoyed by "Known" candidates. Second, the change in support from a candidate's party voters reflects the establishment of a brand name—what Geer (and others) mean by "known." A brand name may be a necessary condition for winning the nomination, but having one gives no indication whether it is a good or a bad brand name—one that will lead the candidate to win the nomination (as Carter and McGovern did) or one that can lead the candidate to win the presidency (as only Carter was able to). Further, once a candidate becomes known, his support might actually decline (as it did for Bush and Hart). As Popkin puts it,

> As voters learn more about a candidate, or as they discover that they know less than they thought, they sometimes learn that they prefer a different candidate. The rapid decline of early "surge" candidates occurs because of new information about their stands or competence, not because of changes in voter expectations about the candidates' chances of electoral success.[90]

On two related claims—that TV picks the winners and that the media is biased—I have not been able to locate any evidence at all. Even if these claims were true, it is still not clear what might have changed from the previous era.

It may be true that the average individual gets most of his information from television, but it does not follow that voters have no other sources of information. A majority of voters read newspapers and, as Popkin argues, people gather information from everyday life and experience.[91] It may also be true that most TV reporting on campaigns covers horse-race and hoopla, especially late in the campaign, but this tells us nothing about the inferences that voters are able to make from such coverage. As for momentum, it tells us nothing about campaign dynamics. In his essay, Aldrich sketches the beginning of a more coherent theoretical approach to the concept, but there is no evidence for the common inferences drawn about the ability of TV to affect campaign outcomes.

(8) *The campaign reforms of 1972 and later mattered.* I have summarized evidence that the reforms (1) helped to bury bosses, and (2) made conventions more demographically representative of the population as a whole. Have these changes created persistent factions? In the days when party bosses ruled with an iron grip, intramachine factions were unthinkable. Factions in the national party, on the other hand, composed of groups of state and local machines, undoubtedly existed but were not easily discerned in any quantitative data. Insofar as party bosses fade into irrelevance, then, we might expect to see signs of policy-based, factional infighting within the parties. And, in fact, Reiter shows that both parties do indeed suffer persistent factionalism (measured by looking at the degree of unity of state delegations on roll-call votes at the national party conventions).[92] However, this factionalism is something that arose in the early 1960s, prior to the McGovern-Fraser reforms.

Have the reforms helped amateurs, opinion outliers, and interest groups take over the nominating conventions? It is unclear whether this is a recent phenomenon. Jeane Kirkpatrick interviewed convention delegates at both major party conventions in 1972; this information was combined with larger surveys to compare Democratic rank-and-file attitudes with convention delegates' attitudes on a range of issues (including welfare, busing, crime, civil rights, and so forth). She found that ordinary Democrats (not convention delegates) were more in tune with delegates to the Republican convention than with delegates to the Democratic convention.[93] This was the case for most issues and candidates, as well as in the aggregate; moreover, the differences were greatest between ordinary Democrats and McGovern delegates!

There is widespread agreement that convention delegates are unrepresentative in one way or another of the whole electorate (for each party as well as for both). Polsby notes that as a group the 1972 Democratic convention delegates were better educated and wealthier than Americans in general.[94] John Kessel agrees, looking at citizens and activists in 1988, and also provides data showing that delegates are unrepresentative of the general electorate in terms of religion, race, sex, and self-identification as conservative, liberal, or moderate.[95] Interestingly, his data show that while Republican delegates are wealthier than Republican identifiers (and also Democratic delegates), on most measures they appear to be closer to the Republican rank and file than are Democratic delegates to theirs. No significance tests are offered.

Geer argues that both Democratic and Republican voters in primaries are more moderate than their party's rank and file.[96] Assuming that Democratic convention delegates have become less representative of the average Democratic voter, have these delegates nominated outliers? Reiter claims, on the contrary, that for both parties "moderate candidates have often defeated more 'extreme' challengers for the nomination—Carter in 1976 and 1980, Ford in 1976, Mondale in 1984." [97] How should one judge how "extreme" a candidate is? The evidence cited both for and against such claims tends to be tautological: so-and-so lost, so he must have been an extremist.

In summary, I find little evidence one way or the other on the effects of the post-1968 party reforms. While convention delegations seem to have become more demographically representative, they apparently have had little or no impact on nominations, especially since most are now committed to a candidate, at least on the first ballot, via a state primary election.

When True Is False: Questioning Premises

On the whole, I have two responses to the stylized facts. I agree that there is some evidence to support the claims that party machines have declined, presidential campaigns have become candidate-centered, voter participation has declined, convention delegates have become more demographically representative of the whole population, and voters get most of their news from television. However, many of these facts were true long before the rise of television. The one change that seems most likely to have arisen from the reforms—the changed demographic composition of party delegations to the convention—poses no identifiable threat to democracy. As for the balance of the claims, I simply do not believe that they stand up to scrutiny.

I have tried to piece together arguments for the stylized facts found in the literature and expounded by political insiders, using premises that are plausible or at least not obviously false. However, the following assumptions—important pieces of the arguments for the stylized facts—have many strikes against them:

(1) The assumption that average demand for information is below the point at which print media become cheaper to consume than watching additional TV is patently silly. Millions of Americans buy

daily newspapers; further, millions listen to the radio and millions more discuss politics with friends, neighbors, and acquaintances at the drop of a hat. Individuals have ready access to many sources of information in addition to the nightly news.

(2) TV networks also spend considerable effort trying to segment their markets in order to provide more targeted audiences for advertisers, contrary to the assumption that they always target the average viewer in general.

(3) The common definition of information as encyclopedic knowledge is clearly flawed. People are capable of making choices in the real world without being "fully informed" in the sense used by economists and game theorists in formal models. Further, we can actually observe individuals making choices between different levels of consumption of media sources of information. It follows that rational individuals are trying to use their resources efficiently, making trade-offs between using resources to collect information about alternative courses of action and to really take action. To be informed means having the ability to make accurate inferences about the world based on a stock of knowledge; it is neither necessary nor sufficient to have encyclopedic knowledge of facts in order to make good inferences.

(4) The assumption that television and candidates manipulate voters implies that people make no effort to predict how their choices will interact with those made by other people to produce outcomes. This is clearly false with regard to predictions we make about what is likely to happen in our everyday lives; the choice of when to brave the commute home from work is a trivial example. Do we leave right at five o'clock and get snarled in rush-hour traffic, or do we stay at the office an extra forty-five minutes, waiting for traffic to clear? The optimal choice before the fact, of course, depends on the choices we expect other commuters to make, since if enough of us choose to wait, traffic will be clear at 5:00 but snarled at 5:45.

(5) The set of individuals' most preferred policy outcomes is uniformly distributed in some interval on a left-right continuum. It is well known that many issues, such as abortion, don't fit easily on a left-right continuum. But even more implausible than that claim is the assertion of a uniform distribution of preferences. Every schoolchild knows, with good reason, that when we don't know anything about a population, it can be useful to assume that it is distributed normally (a bell curve). We tend to believe that there is probably some average or central tendency in the population, be it height, weight, intelligence, or opinion. Why, then, without any evidence, should we expect voters to be different? But without a flat distribution of preferences, it is difficult to sustain arguments that extremist interests can take over political conventions.

The implausibility of these assumptions and relying on them for any internally consistent theory of presidential nominations that purports to explain the stylized facts should be sufficient to dispense with most of the stylizations I have discussed. So what can we conclude from the literature on presidential nominations and the role of television and of the post-1968 party reforms? We can still believe that the media have changed since the 1960s and Watergate. We can believe that the structure and organization of campaigns have also changed as candidates and their hired guns learn how best to present their message to the voters, given the resource constraints and the availability of free and paid media. And we can believe that voters' expectations and demands have changed as new information technologies and campaign strategies have made more and different information available to them. The following essays present detailed arguments and evidence in support of these circumstances that we believe to be true. I have not, however, found any evidence that these changes have affected the quality of American democracy in any significant way.

Notes

1. Nelson W. Polsby and Aaron B. Wildavsky, *Presidential Elections: Strategies of American Electoral Politics*, 2nd ed. (New York: Charles Scribner's Sons, 1968), 28; see also Polsby and Wildavsky, *Presidential Elections: Contemporary Strategies of American Electoral Politics* (New York: Charles Scribner's Sons, 1964); Austin Ranney, "The Democratic Party's Delegate Selection Reforms, 1968-76," in *America in the Seventies*, ed. Allan P. Sindler (Boston: Little, Brown, 1977), 204; Everett Carll Ladd, Jr., *Where Have All the Voters Gone? The Fracturing of America's Political Parties* (New York: W. W. Norton, 1977), 52.
2. Polsby and Wildavsky, *Presidential Elections,* 1968, 78.
3. Theodore White, *The Making of the President, 1968* (New York: Pocket Books, 1970), 82; Gary R. Orren and William G. Mayer, "The Press, Political Parties, and the Public-Private Balance in Elections," in *The Parties Respond: Changes in the American Party System,* ed. L. Sandy Maisel (Boulder, Colo.: Westview Press, 1990), 210.
4. White, *The Making of the President, 1968,* suggests that the threshold had been crossed by the election of 1968.
5. Nelson W. Polsby and Aaron B. Wildavsky, *Presidential Elections: Contemporary Strategies of American Electoral Politics,* 7th ed. (New York: The Free Press, 1988), 260-261; William J. Crotty, *Decisions for the Democrats: Reforming the Party Structure* (Baltimore: Johns Hopkins University Press, 1978), 267.
6. Nelson W. Polsby, *Consequences of Party Reform* (New York: Oxford University Press, 1983), 11, 13-14.

7. Theodore White, *The Making of the President, 1960* (New York: Atheneum, 1980), 79, 96-114; William Crotty and John S. Jackson III, *Presidential Primaries and Nominations* (Washington, D.C.: CQ Press, 1985), 15, 17; Polsby and Wildavsky, *Presidential Elections*, 1968, 83; William R. Keech and Donald R. Matthews, *The Party's Choice* (Washington, D.C.: Brookings Institution, 1976), 121-123; Polsby, *Consequences of Party Reform*, 15; Larry M. Bartels, *Presidential Primaries and the Dynamics of Social Choice* (Princeton: Princeton University Press, 1988), 15.

8. John G. Geer, *Nominating Presidents: An Evaluation of Voters and Primaries* (Westport, Conn.: Greenwood Press, 1989), 2.

9. Polsby, *Consequences of Party Reform*, 170.

10. Polsby and Wildavsky, *Presidential Elections*, 1964, 64; see also Howard L. Reiter, "The Limitations of Reform: Changes in the Presidential Nominating Process," Essex Papers in Politics and Government, no. 20 (Department of Government, University of Essex, Wivenhoe Park, Colchester, England, 1984), 4-5; Samuel L. Popkin, *The Reasoning Voter* (Chicago: University of Chicago Press, 1991), 221-222.

11. Martin P. Wattenberg, *The Decline of American Political Parties, 1952-1988* (Cambridge, Mass.: Harvard University Press, 1990), 143-144; see also Gerald M. Pomper and Susan S. Lederman, *Elections in America: Control and Influence in Democratic Politics*, 2nd ed. (New York: Longman, 1980); Angus Campbell, Philip E. Converse, Warren E. Miller, and Donald E. Stokes, *The American Voter* (Chicago: University of Chicago Press, 1960).

12. Quoted in Wattenberg, *The Decline of American Political Parties, 1952-1988*, 149.

13. White, *The Making of the President, 1960*, 82-83.

14. See, e.g., Howard L. Reiter, *Selecting the President: The Nominating Process in Transition* (Philadelphia: University of Pennsylvania Press, 1985), 10; Thomas R. Marshall, *Presidential Nominations in a Reform Age* (New York: Praeger, 1981), 57; Ladd, *Where Have All the Voters Gone?*, 52; Ranney, *America in the Seventies*, 204; Martin P. Wattenberg, "Participants in the Nominating Process: The Role of the Parties," in *Before Nomination: Our Primary Problems*, ed. George Grassmuck (Washington, D.C.: American Enterprise Institute, 1985), 50; and Jeane Kirkpatrick, *Dismantling the Parties: Reflections on Party Reform and Party Decline* (Washington, D.C.: American Enterprise Institute, 1978), 5-6.

15. See, e.g., Stephen J. Wayne, *The Road to the White House*, 4th ed. (New York: St. Martin's Press, 1992).

16. Reiter, *Selecting the President*, 35, Table 2.5.

17. Polsby, *Consequences of Party Reform*, 73; see also Wayne, *The Road to the White House*, 28; Stephen Hess, *The Presidential Campaign* (Washington, D.C.: Brookings Institution, 1988), 47.

18. See, e.g., Larry Sabato, *The Rise of Political Consultants* (New York:

Basic Books, 1981); Melvyn H. Bloom, *Public Relations and Presidential Campaigns: A Crisis in Democracy* (New York: Thomas Y. Crowell, 1973), 250-251.

19. Polsby, *Consequences of Party Reform*, 73.
20. Herb Asher, "The Three Campaigns for President," in *Presidential Selection*, ed. Alexander Heard and Michael Nelson (Durham, N.C.: Duke University Press, 1987), 226.
21. Polsby and Wildavsky, *Presidential Elections*, 40.
22. Wattenberg, *Decline of American Political Parties, 1952-1988*, Table 3.1., 42.
23. Reiter, *Selecting the President*, 7-8.
24. Polsby, *Consequences of Party Reform*, 62.
25. Wayne, *The Road to the White House*, 133; see also Bartels, *Presidential Primaries and the Dynamics of Social Choice;* and John Aldrich, "Methods and Actors," in *Presidential Selection*, ed. Alexander Heard and Michael Nelson (Durham, N.C.: Duke University Press, 1987), 177-180.
26. David S. Broder, "The Parties: Democrats," in David Broder [et al., the staff of the *Washington Post*], *The Pursuit of the Presidency* (New York: Berkeley Books, 1980), 192-194.
27. Jeane J. Kirkpatrick, *The New Presidential Elite: Men and Women in National Politics* (New York: Russell Sage Foundation, 1976); Kirkpatrick, *Dismantling the Parties*.
28. Everett Carll Ladd, *Where Have All the Voters Gone?* 37, 65; see also Samuel Kernell, *Going Public: New Strategies of Presidential Leadership* (Washington, D.C.: CQ Press, 1986), 39-40; Broder, "The Parties: Democrats"; Polsby and Wildavsky, *Presidential Elections*, iix; Byron E. Shafer, *Quiet Revolution: The Struggle for the Democratic Party and the Shaping of Post-Reform Politics* (New York: Russell Sage Foundation, 1983).
29. Polsby and Wildavsky, *Presidential Elections*, iix; see also Aldrich, *Presidential Selection*, 184; Keech and Matthews, *The Party's Choice;* Geer, *Nominating Presidents*, 7; Kirkpatrick, *Dismantling the Parties*, 7.
30. Richard Watson, *The Presidential Contest*, 2nd ed. (New York: John Wiley and Sons, 1984), 98.
31. See, e.g., Crotty, *Decision for the Democrats*, 255-257; Kirkpatrick, *The New Presidential Elite*, 365; Kirkpatrick, *Dismantling the Parties*, 8.
32. Quoted in Polsby, *Consequences of Party Reform*, 34. For a detailed account of the history of the commission, and its predecessor, the Hughes Commission, see Byron Shafer, *The Party Reformed: Reform Politics in the Democratic Party 1968-1972* (New York: Russell Sage Foundation, 1979).
33. Jeffrey B. Abramson, F. Christopher Arterton, and Gary R. Orren, *The Electronic Commonwealth* (New York: Basic Books, 1988).
34. Marshall, *Presidential Nominations in a Reform Age*, 56.
35. Robert E. DiClerico and Eric M. Uslaner, *Few Are Chosen: Problems in*

Presidential Selection (New York: McGraw-Hill, 1984), 55, Table 2.2.
36. Sabato, *The Rise of Political Consultants;* see also Chapter 4 in this volume.
37. Aldrich, *Before the Convention,* 64, quoting Michael Robinson, 65; see also conference transcripts; Geer, *Nominating Presidents,* 5-6.
38. Aldrich, *Before the Convention,* 118; see also, DiClerico and Uslaner, *Few Are Chosen,* 69; Benjamin I. Page and Mark P. Petracca, *The American Presidency* (New York: McGraw-Hill, 1983), 102; Bartels, *Presidential Primaries and the Dynamics of Social Choice;* James Davis, *Presidential Primaries: The Road to the White House* (Westport, Conn.: Greenwood Press, 1980), 82; James R. Beninger, "Polls and Primaries," in *Presidential Primaries: The Road to the White House,* ed. James W. Davis (Westport, Conn.: Greenwood Press, 1980), 115; Wayne, *The Road to the White House,* 118; John Kessel, *Presidential Campaign Politics,* 3rd ed. (Chicago: Dorsey Press, 1988), 9; Marshall, *Presidential Nominations in a Reform Age,* 57; Paul R. Abramson, John H. Aldrich, and David W. Rohde, *Change and Continuity in the 1980 Elections* (Washington, D.C.: CQ Press, 1982), 28; Paul R. Abramson, John H. Aldrich, and David W. Rohde, *Change and Continuity in the 1984 Elections,* rev. ed. (Washington, D.C.: CQ Press, 1987), 22; Paul R. Abramson, John H. Aldrich, and David W. Rohde, *Change and Continuity in the 1988 Elections* (Washington, D.C.: CQ Press, 1990), 18, 31, 37; William G. Mayer, "The New Hampshire Primary: A Historical Overview," in *Media and Momentum: The New Hampshire Primary and Nomination Politics,* ed. Gary R. Orren and Nelson W. Polsby (Chatham, N.J.: Chatham House, 1987); Henry E. Brady and Richard Johnston, "What's the Primary Message: Horse Race or Issue Journalism?" in *Media and Momentum: The New Hampshire Primary and Nomination Politics;* Keech and Matthews, *The Party's Choice,* 95-96; Watson, *The Presidential Contest,* 42.
39. Iyengar and Kinder discuss the vulnerability of typical individuals' opinions to subtle forms of manipulation of the news, as might follow from a mildly ideologically biased newscast. Shanto Iyengar and Donald Kinder, *News That Matters* (Chicago: University of Chicago Press, 1987).
40. See, e.g., Sabato's essay in this volume.
41. Reiter, *Selecting the President,* 2. Subsequent reform efforts reinforced the general thrust of McGovern-Fraser. The Mikulski Commission, for example, "tightened the proportional representation rules for the 1976 convention, allegedly destroying the ability of party leaders to bring to conventions delegations that are united under their control. . . . Other significant Democratic efforts have included the . . . Winograd Commission, which mandated that half the delegates to the 1980 convention be women, and the . . . Hunt Commission, which increased the participation of party and public officials and relaxed some of the earlier reforms in anticipation of the 1984 convention." Reiter, *Selecting the President,* 4. Similarly, the 1974 Federal Election Campaign Act. Then came the

FECA rules: matching funds "were expected to affect the nominating process . . . by making it easier for candidates of modest means to pay for their campaigns and to avoid becoming captives of wealthy contributors and special interests." Reiter, *Selecting the President*, 4; for a full exposition of the post-1968 reforms, their formulation, and their adoption, see Shafer, *Quiet Revolution;* Polsby, *Consequences of Party Reform,* 40-52; see also Reiter, "The Limitations of Reform"; James W. Ceasar, *Reforming the Reforms* (Cambridge, Mass.: Ballinger, 1982), 31-54.

42. Reiter, *Selecting the President*, 12.

43. Ranney, *America in the Seventies*, 204.

44. Reiter, *Selecting the President*, 3-4.

45. Polsby, *Consequences of Party Reform*, Table 1.1., 11; Reiter, *Selecting the President*, Table 3.2, 48. The story in 1968 is complicated by Johnson's withdrawal from the race after his near-defeat at the hands of Eugene McCarthy in New Hampshire and by Robert Kennedy's assassination following his victory in the California primary. Kennedy won five primaries and 30.6 percent of the votes cast; McCarthy won six and 38.7 percent respectively. Humphrey, the sitting vice president, did not contest any primaries. See Polsby, *Consequences of Party Reform,* 26, note 54.

46. Reiter, *Selecting the President,* Table 3.1, 44.

47. Reiter, *Selecting the President,* Table 4.8, 74.

48. Ranney, *America in the Seventies*, 1977.

49. Polsby, *Consequences of Party Reform;* Reiter, *Selecting the President,* Chapter 5; Anne Costain, "An Analysis of Voting in American National Nominating Conventions, 1940-1976," in *American Politics Quarterly* 6 (January 1978): 95-120.

50. James Ceasar, *Presidential Selection: Theory and Development* (Princeton, N.J.: Princeton University Press, 1979), 300.

51. For a discussion see Reiter, *Selecting the President,* 136-139.

52. Crotty, *Decision for the Democrats;* Xandra Kayden, "Regulating Campaign Finance: Consequences for Interests and Institutions," in *Presidential Selection,* ed. Alexander Heard and Michael Nelson (Durham, N.C.: Duke University Press, 1987); John Kessel, *Presidential Campaign Politics: Coalition Strategies and Citizen Response* (Homewood, Ill.: Dorsey Press, 1980); Polsby, *Consequences of Party Reform.*

53. On framing and priming, see Iyengar and Kinder, *News That Matters.* I do not mean to imply, nor should the reader infer, that candidates' attempts to manipulate voters are necessarily successful. Though the claim that candidates manipulate voters is not a prominent one in the literature, it is occasionally made or implied in some arguments. If we want to believe that voters *are* manipulable, we must also assume, first, that people are unsophisticated (unable to do backward induction), and, second, that people have no prior information or beliefs. If these two assumptions are true (a controversial point, to be sure), then we can deduce that candidates can manipulate voters.

54. Crouse offers support for the view that better-known, "establishment"

candidates get the lion's share of press coverage. Timothy Crouse, *The Boys on the Bus* (New York: Ballantine Books, 1973).

55. Actually this applies to *potential* primary voters. We might expect those who care enough to go out and vote in primaries to expend resources on obtaining information about candidates from sources other than TV.

56. See, e.g., Polsby, *Consequences of Party Reform*.

57. Arthur W. Lupia, "Busy Voters, Agenda Control and the Power of Information," *American Political Science Review* (June 1992).

58. John R. Petrocik, "The Theory of Issue Ownership: Issues, Agendas, and Electoral Coalitions," 1991, under submission.

59. See Gary Jacobson, *The Politics of Congressional Elections*, 3rd ed. (New York: HarperCollins, 1992).

60. This, of course, depends crucially on the relative prices of different mass media advertising. I have assumed certain things about the market for information and consumer demand that imply that the average consumer will get all of his or her information from television. In the simple model, with only two competing TV stations, both targeting the average viewer, the equilibrium price for TV ads, denominated as dollars per viewer-second, will be determined trivially by the intersection of the aggregate supply curve with the aggregate advertiser demand curve.

61. Geer, *Nominating Presidents*, 80-81.

62. Indeed, if it serves the group's purpose, it will subsidize the provision of such information to nongroup members.

63. People in groups are better informed than people not in groups, all else being equal. See Phillip E. Converse, "The Nature of Belief Systems in Mass Publics," in *Ideology and Discontent*, ed. David Apter (London: Free Press of Glencoe, 1964), 236, Figure 3.

64. Thomas Schwartz, "Your Vote Counts on Account of the Way It Is Counted: An Institutional Solution to the Paradox of Not Voting," University of Texas, Austin, 1986, unpublished paper; see also Mancur Olson, *The Logic of Collective Action* (Cambridge, Mass.: Harvard University Press, 1965); Sidney Verba, Norman H. Nie, and Jae-On Kim, *Participation and Political Equality: A Seven-Nation Comparison*, ed. George Grassmuck (Washington, D.C.: American Enterprise Institute, 1978), 301-302.

65. See, e.g., Crotty and Jackson, *Presidential Primaries and Nominations*, 23; Reiter, *Selecting the President*, Tables 3.1, 3.2.

66. Paul T. David, Ralph M. Goldman, and Richard C. Bain, *The Politics of National Party Conventions* (Washington, D.C.: Brookings Institution, 1960), quoted in Reiter, *Selecting the President*, 10.

67. See Harold W. Stanley and Richard A. Niemi, *Vital Statistics on American Politics* (Washington, D.C.: CQ Press, 1988).

68. See Mathew D. McCubbins and Frances Rosenbluth, "Electoral Structure and the Organization of Policymaking in Japan," University of California, San Diego, 1991, unpublished; see also McCubbins and Rosenbluth, "The Liberal Democrats Face Japan's New World," *The*

Economist 314 (24 February 1990): 31.
69. White, *The Making of the President, 1960.*
70. Sabato, *The Rise of Political Consultants,* Appendix C.
71. Keech and Matthews, *The Party's Choice,* 137.
72. Wattenberg, *The Decline of American Political Parties,* 92-98, and figures 6.1-6.3.
73. These statistics were calculated from data presented in Walter Dean Burnham, "The Turnout Problem," in *Classic Readings in American Politics,* 2nd ed., ed. Pietro S. Nivola and David H. Rosenbloom (New York: St. Martin's Press, 1990), Table 18, 136-137; and Abramson, Aldrich, and Rohde, *Change and Continuity in the 1988 Elections,* 91.
74. Under the assumption that data are drawn from random samples of the voting-age population. See Wattenberg, *The Decline of American Political Parties,* Table 3.1, 42.
75. Gary C. Jacobson, *The Electoral Origins of Divided Government: Competition in U.S. Elections, 1946-1988* (Boulder, Colo.: Westview Press, 1990), 9; see also Martin P. Wattenberg, *The Decline of American Political Parties, 1952-1988.*
76. See Stanley and Niemi, *Vital Statistics on American Politics,* 129.
77. See Reiter, *Selecting the President,* 28, Table 2.2.
78. Reiter, *Selecting the President,* 27, Tables 2.3, 2.4.
79. Reiter, "The Limitations of Reform," 8-9.
80. Table 2-13, Stanley and Niemi, *Vital Statistics on American Politics,* 69.
81. Table 2-12, Stanley and Niemi, *Vital Statistics on American Politics,* 67-68.
82. Brady and Johnston, *Media and Momentum,* 163, referring to Table 5.9.
83. Table 2-6, Stanley and Niemi, *Vital Statistics on American Politics,* 57.
84. Popkin, *The Reasoning Voter,* 23-23.
85. Anthony Downs, *An Economic Theory of Democracy* (New York: Harper & Row, 1957), 85.
86. Lupia, "Busy Voters, Agenda Control and the Power of Information."
87. Jacobson, *The Electoral Origins of Divided Government.*
88. See, e.g., Popkin, *The Reasoning Voter,* 57, Table 3.1; Jacobson, *The Electoral Origins of Divided Government,* 112-115; Martin Wattenberg, *The Decline of American Political Parties, 1952-1984* (Cambridge, Mass.: Harvard University Press, 1986), Tables 9.8-9.11.
89. Geer, *Nominating Presidents,* 81-82, figures 5.1, 5.2 show that momentum does play a role for relatively unknown candidates. In such cases, a large part of a candidate's support appears to depend on his or her perceived ability to win votes. But as candidates become more well known—and for candidates who are not ciphers from the outset—issue stances and character dominate people's perceptions of them. Also, if we examine Aldrich's tables 6-7 and 16 comparing delegates won from February to June of election years with the amount of money raised each of those months, we find no significant correlation (even with a lag structure) between winning delegates and raising money.

90. Popkin, *The Reasoning Voter,* 136.
91. Popkin, *The Reasoning Voter.*
92. Reiter, *Selecting the President,* Chapter 5.
93. Kirkpatrick, "Representation in American National Conventions," Table 9, 371; see also tables 10, 11.
94. Polsby, *Consequences of Party Reform,* 161, Table 5.2.
95. John Kessel, *Presidential Campaign Politics,* 4th ed. (Pacific Grove, Calif.: Brooks/Cole, 1992), Table 3-1, 80; see also Reiter, *Selecting the President,* tables 4.1-4.6.
96. Geer, *Nominating Presidents,* Chapter 2, Tables 2.1-2.3.
97. Geer, *Nominating Presidents,* 26.

2. PRESIDENTIAL CAMPAIGNS IN PARTY- AND CANDIDATE-CENTERED ERAS

John H. Aldrich

"Prior to the advent of television," said political consultant Joseph Napolitan, "the message from the candidate had to be filtered either through a party or through an editor or a news editor. It never got directly from the candidate to the people. With television, we could begin to communicate directly to the people in their living room."

John Aldrich argues that the advent of direct communication between candidates and voters led to a new kind of campaigning. Issues take a back seat to the quest for good press reports—however empty of content—and the candidates' desire to squelch unfavorable coverage of their records and personal lives. Issues, after all, are not news; events are news. So, what candidates say becomes less important than how and where they say it. As University of California at San Diego professor Dan Hallin noted, "Rather than simply showing you what the candidate said, the journalist is telling you the story about this event and its significance, giving you an analysis of it. And then the campaign consultants and other experts are brought in to analyze that."

Aldrich says that modern presidential candidates concentrate on refining their images at the expense of defining substantive policy positions. Campaign organizations therefore need and attract workers who focus on the candidate's image, not the candidate's party or policy positions. The ubiquitous presence of television, together with these new candidate-centered campaigns, has given rise to a new class of campaign workers and professionals: the vassals of the new feudalism, who rise and fall with the triumphs and failures of single candidacies instead of the broader fortunes of parties and party machines. —Ed.

The subject of this essay is the organization and design of presidential campaigns. In many ways, to think of a presidential campaign as organized is, if not an oxymoron, at least a non sequitur. Will Rogers's oft-quoted line, "I belong to no organized party. I am a Democrat," seems to be an even more apt characterization of a presidential campaign organization than a major, nearly two-century-old political party. In part, this apparent disorganization is due to the basic nature of campaign organizations, notably their necessarily transient existence.

And yet campaigns may indeed be highly organized. They start, for example, with one great advantage over other political organizations such as the U.S. Congress. A presidential campaign organization has a very clearly and narrowly defined objective—winning the nomination and then the presidency for its candidate—and as a result it is naturally inclined to be more orderly, if not necessarily organized, than a Congress composed of 535 individuals with 535 objectives (or perhaps a shared objective such as reelection, realized in 535 different ways). Congress, moreover, deliberates in a complex way, culminating in decision by majority rule whose complications rarely occur in campaign decision making.

There is a conventional understanding that at least some campaigns are well organized. Thus it is understood that Michael S. Dukakis was advantaged among the "seven dwarfs" running for the 1988 Democratic nomination (as contenders other than front-runner Gary Hart were called) by virtue of his stronger and more effective organization. On the other side in 1988, George Bush benefited from a strong and effective organization in both spring and fall. Assuming there is some validity to these beliefs, the immediate inference is that organizational capacity matters. The person who has been best organized wins, or at least is greatly advantaged. Of course, this may be the result of hindsight. That is, an organization may seem to have been effective because its candidate won, and winning, on its own, is likely to improve organizational effectiveness. Put alternatively, losing (a primary or caucus or, in the fall, declining poll standings) puts strains on a campaign organization that weaken, at times irreversibly, its effectiveness. Did Dukakis's general election campaign seem more poorly organized than in the spring because it could not make the transition to the needs of a general election, or because, like all campaign organizations, it was brittle and cracked under stress when serious problems arose? The answer is probably some of each.

In what follows, I look at the change from party-centered to candidate-centered elections, and how this massive set of changes has affected the nature and design of presidential campaign organizations. In particular, I argue that presidential campaigns in party-centered elections, such as in the "golden age of parties" (roughly, from the end of Reconstruction until at least the Progressive era), can be seen as analogous to "feudal" campaigns. These were rooted in local party organizations, and thus in the geography that defines electoral districts, just as feudalism was rooted in land. Contemporary campaigns in candidate-centered elections are like a modified form of feudalism, called "bastard feudalism," rooted in personal ambitions as well as in the earlier feudal basis of party and, to an even lesser extent, electoral geography. Then I examine the current arrangements typical of presidential campaigns, incentives that structure their organization, and consequences of those incentives and organizations for governance when the winning candidate takes office.

Presidential Campaigns and Bastard Feudalism

Lewis Chester, Godfrey Hodgson, and Bruce Page, three reporters for the *Sunday Times* of London, wrote a wonderful book on the 1968 presidential campaign.[1] Studying the entourages surrounding the candidacies of Senator Robert F. Kennedy and Governor Nelson A. Rockefeller in 1968, they made an analogy to fourteenth-century "bastard feudalism," as explicated by the Oxford don K. B. McFarlane. Money began to replace land as the basis for the connection between lord and vassal in that century. As a result, allegiances were more temporary than before and vassals "began to choose as their patron the man who could do them the most good" (p. 211). They quoted McFarlane as writing:

When a man asked another to be his "good lord" he was acquiring a temporary patron. In this loosely-knit and shamelessly competitive society, it was the ambition of every thrusting young gentleman—and also of everyone who aspired to gentility—to attach himself for as long as suited him to such as were in a position to further his interests.... [Attached to the "good lord"] there accumulated a vast but indefinite mass of councillors, retainers, and servants, tailing off into those believed to be well-wishers.

McFarlane called these hangers-on "bastard feudatories" and the whole system "bastard feudalism" (p. 211).

The authors then pointed to the Kennedy and Rockefeller families as leading examples of "good lords." In particular, they suggested that the old politics was more like feudalism, where party and locale determined loyalties rather like land did in feudal times. In the new politics, loyalties were based to a greater extent on the ambition of individuals who aspired to political influence through a candidate. Such ambitious types attach themselves to a "good lord" such as Kennedy or Rockefeller, but only for as long as it suits such politicos and only for as long as both sides are in a position to further their own interests. Thus, more so than in the past, key advisers are for hire, as opposed to the time when a candidate's campaign was more heavily determined by the extant party apparatus. In extreme cases, even party affiliation can be violated in new bastard feudal campaigns. Thus advisers gravitate to the strongest and most influential politicos, where a combination of money and potential political influence characterize both contemporary campaigns and bastard feudalism. To be sure, there are limits—party affiliation is rarely violated—and there are some, such as Hamilton Jordan, adviser to Jimmy Carter, whose political loyalty has been akin to the old feudal bonds rather than to bastard feudalism. These limits explain why the analogy is to bastard feudatories rather than to mercenaries, that is, to those motivated exclusively by personal ambition.

Modern presidential campaign organizations (since about 1960) reflect the shift from party-centered to candidate-centered campaigns. An individual candidate can and does build a personal campaign organization rather than adopting in whole or very large part the extant party apparatus of state and local committees and organizations. The attraction to the candidate is based in part on loyalties of various sorts (partisan, ideological, policy, even personal ties that transcend politics), as well as on the perception of those who seek to, or can be convinced to, join the campaign that the candidate will be likely to further their personal ambitions. Some candidates, such as the Kennedys and Rockefellers, have such substantial resources, political influence, and the like that they are able to attract the "best and the brightest," and are thereby advantaged. Others may begin their quest for nomination with less to attract bastard feudatories, but as success grows, their chances of winning increase, as does their appeal to the ambitious. Conversely, if a campaign falters, especially in the race for the presidential nomination, the candidates lose momentum. This means not just a loss of public support but even more a loss of organizational capacity and resources to campaign effectively. The personal ambitions of those who could further the campaign organization are no longer well served by a candidate who has lost momentum.

Whence Cometh the Candidate-Centered Campaign?

Political science has developed more than a mere cottage industry to explain the rise of candidate-centered campaigns and the concomitant decline of political parties. Voters have become less strongly attached to either political party, and the incidence of self-proclaimed independence has risen. There is a widespread belief that parties are no longer as relevant to elections or to governance. Some argue that the decline is less a function of alienation from parties than it is simple indifference to and perceived irrelevance of parties.[2] Incumbency per se has come to rival partisanship as the most important determinant of voters' choices in congressional elections.[3] Incumbents invariably seem to win if they avoid personal scandal, and they have become able to develop personal campaign organizations, raise vast sums of money, and structure information about themselves.[4] They, not parties, structure congressional campaigns, just as presidential contenders structure their campaigns. Voters split their tickets with increasing regularity. The result is divided government, with Republicans apparently having a lock on the presidency while Democrats are even more firmly in control of the House and perhaps the Senate. State governments are just as likely to be under divided party control, with voters apparently splitting their tickets at the state and local levels as well.[5] Perhaps, as syndicated columnist David Broder put it in the book that helped launched this disciplinary industry, *The Party's Over*.[6]

Some place the responsibility for the decline of parties in the parties' own hands. The adoption of the McGovern-Fraser Commission reforms at the 1972 Democratic national convention, along with enactment of federal campaign finance reforms, led to substantial changes in the presidential nomination process, notably the rapid expansion in the use of primary elections.[7] As a result, presidential candidates *had* to conduct large-scale public contests for nomination, and these all but necessarily *had* to be independent of traditional party organizations. As historians have long told us, parties were first and foremost concerned with the election of presidents at the national level.[8] All else might be local, but party politics at the national level was presidential politics. Party leaders ceded to others the control of presidential nominations, and thus the center of national parties. And in part, they chose to do so themselves.

But this explanation cannot be right. Broder, after all, told us that the party was over before McGovern-Fraser reforms had their impact and before campaign finance reforms were enacted. Bastard feudal campaigns were perceived in 1968, and extended back to 1960, as the above-cited examples indicate. More to the point, virtually all measures of party decline, decay, and decomposition were in place and under way in the mid-1960s; many ended their steepest decline around 1972. The public's party loyalties declined in or around 1966. Incumbency as a rising force in congressional campaigns dates back to 1966. Divided government began in January 1969. The Republican lock on the presidency, if there is one, dates from Nixon's election, and of course the Democrats have controlled the House continuously since 1954.[9]

The changes that some view as so dramatic and as boding so ill for the republic coincided with the first appearance of bastard feudalism. They occurred over a long period, to be sure, but reached a culmination in the mid-1960s. What are the sources of these changes for presidential campaign organizations in particular? They are due to changes in the environment external to presidential campaigns. In part, they are due to the decline of political parties, but the irrelevance of parties caused a relatively consistent decline that began in the 1880s.[10] These changes did not *greatly* affect presidential campaigns. More accurately, they did not greatly affect presidential campaigns until combined with a second, and more rapid, set of changes: technological transformations that made it feasible to run a personal, candidate-centered campaign for nomination and election.

Two "Textbook" Versions of Political Parties and Their Typical Presidential Campaign Organizations

Each presidential campaign is unique in important ways. Every four years a new and different set of organizations is created for the campaign. I believe, however, that there are important differences between the sorts of campaign organizations employed earlier in our history and those of today. The change was gradual and erratic, but it is fair to contrast a typical organization from the era of party-centered elections and one from today's candidate-centered contests. The differences, I believe, are due to the changing nature and role of political parties. I idealize these as "textbook" accounts to emphasize that at no time were parties exactly like either of these two views, but that the party-centered and candidate-centered eras are typified by them.[11]

Campaign Organizations in the Golden Era of Parties

To follow the above analogy, this era corresponds to the feudal age of parties, most nearly approximated in the late nineteenth century. As Richard P. McCormick has demonstrated, the national party was first and foremost a *presidential* party.[12] But, as he also observed, the party was, on a continuing and "most important" basis, a congeries of state and local party organizations.[13] It met as a national party only once every four years. Much business was conducted at the national convention, but the most important national action was the presidential nomination, selection of the party's standard-bearer. Presidential contenders did not directly participate in the proceedings, they did not overtly campaign beforehand, and their business at the convention was conducted by others. There was no observable presidential nomination campaign organization. Instead, party brokers wheeled and dealed in smoke-filled rooms, shifted alliances, cut deals, and offered "their" delegates as best they saw fit. Only later, at the turn of the century, was there an overt presidential campaign organization, and it was primarily a *party* organization for mobilizing voters in the fall. It would be still later before nomination organizations surfaced, and not until the 1960s before such public campaigns were successful.

The glue that held the party together in this era was patronage and the spoils system. Joseph A. Schlesinger distinguishes between two types of partisans: the party's "office seekers" and its "benefit seekers." [14] The first want office in order to realize their own ambitions; the second want to receive the benefits that flow from the control of those offices by others. By and large, candidate and activist depended upon the party rather than governmental office to achieve their goals over the long term. For the ambitious politician, a long career was likely to involve the holding of numerous offices. The careerist would typically rotate in and out of Washington, the statehouse, the courthouse, and/or the city hall. In between, the party would provide sinecures until the next rotation brought the office seeker back to a formal position of power. While differing more in degree than in kind from today, far fewer people made a career out of holding a particular office. The benefits the activist was seeking were patronage based and may have been appointed governmental positions, party offices, or jobs in the private sector. Patronage also included contracts awarded by the victorious party in office, providing money to the company and jobs to its employees.

Governmental, especially national governmental, policy was much more likely to allocate distributive benefits, as Theodore Lowi reminds us.[15] Campaigns were fought over such internal improvements as Clay's "American system" (essentially a platform advocating use of governmental resources for building the nation's infrastructure, such as roads and canals) in the 1830s and 1840s. Industrial development policies were more typical in the late nineteenth century, but these were also distributive policies. Tariff policy was a recurring issue throughout the nineteenth century, and it, too, provided protection for some, development for some, restrictions for others. That is, the government, for the most part, allocated benefits of a rather discrete sort that could be distributed closely along geographical lines (to states, cities, or congressional districts). Being primarily divisible, government benefits could be targeted to geographic units on the basis of elections and, especially, party control.

In short, whether patronage-style jobs and contracts or distributive policies, the action of the party period was in the spoils system. This system of benefits and their distribution required the control of office. Benefit seekers and office seekers alike, that is, profited from the control of office and·lost all, at least temporarily, from the loss of that control. With the careerist ambitions of office seekers, as well as the benefits sought by activists, tied to the good fortunes of the party, both groups shared a very strong incentive to see the party win. Which individuals won or what ideology they espoused was of far less importance than the party's control of office, with its ability to offer outlets for the ambitious politician and to offer the spoils of office to the activist cum benefit seeker.

Just as elections were based on geography, so, too, was the party rooted in geography. This made state and local machines possible and, at the national level, made plausible former House Speaker Tip O'Neill's favorite aphorism that "all politics is local." National appeal mattered only insofar as national candidates were able to stimulate turnout from the top of the ticket and as winning the presidency and Congress made possible the distribution of the spoils of national office through the party. Moreover, votes can easily be counted, so rewards could be dispersed to where the votes came from. In the golden age of parties, the national, that is, the presidential, contests were often very close, as was the proportion of Democrats to Republicans in Congress, but cities, states, even whole regions of the country were solidly Democratic or Republican.

Political actors, whether of the office- or benefit-seeking type, cared primarily about winning, for only control of office mattered to realize their ambitions. And that was rooted geographically. Thus was the party, like the feudal system, tied to the land. One backed one's party above all in interparty competition. One backed the candidate of the region in intraparty competition for much the same reason, because victory meant even more of the spoils of office would flow that way.

Campaigns reflected those incentives. With ambitious politicians and benefits seekers alike tied to party—and to regional party—by the dominating goal of victory, the set of regional parties was thus the natural and inevitable place to build a campaign organization.

Parties in this period held a virtual monopoly on the resources necessary to conduct a credible presidential campaign. Campaigns were labor-intensive devices to locate potential supporters, woo them to the party's side, and mobilize them on election day. In today's parlance, it was "retail" campaigning, seen today at the presidential level only in the Iowa caucuses and New Hampshire primaries, rather than "wholesale" campaigning for votes, such as through television and mass mailings. The availability of labor was made possible by the spoils system; indeed the party labor pool was often referred to as an army mobilized by a fervor analogous to a crusade—backed up by the spoils system and focused on the goal of winning election for as many candidates as the party leadership chose to run. A nominee simply had no effective choice but to rely on the various state and local party organizations—and hence on the promise of distributional goods his victory would bring—to mount a serious general election campaign.

As for nomination campaigns, there was neither the need for nor the technological possibility of running a national public campaign. Moreover, because of coinciding interests, as well as control over the patronage system, bosses could offer or withhold virtually their entire bloc of delegates as they saw fit. Thus a nomination campaign was primarily bargaining among a relatively small number of party chieftains. A candidate's nomination campaign "organization" was a small group of backers, dramatically unlike the contemporary national campaign organization needed to run in the primary season. Such an organization provides a very large part of the contemporary general election campaign organization, whereas in the golden age there was precious little to transfer from nomination to general election.

Suppose, for the moment, that someone sought the presidency outside the two-party system. William Jennings Bryan, for instance, could have run for president in 1896 as a Populist (only). He chose instead to be a Democrat, securing the party's nomination after his dramatic "Cross of Gold" speech to the convention.[16] In today's parlance, his choice reflected the high barriers to entry effectively preventing those lacking the aid of partisan troops and resources from competing successfully even with the support of a third party that has had a fair degree of electoral success. Moreover, had Bryan been elected as a third-party president, his ability to work with either major party in Congress would have been greatly attenuated unless he rewarded local party machines by allocating distributive benefits through Congress— unless, that is, he effectively became a Democratic president. Or consider Theodore Roosevelt in 1912, the most popular politician of his era. That year, for the first time, a significant number of presidential primaries (thirteen) were held. Roosevelt won virtually all of them, but the GOP organization was sufficiently strong to deny him the nomination. Seeing defeat, his supporters left the Republican convention, and he launched a third-party presidential bid. But even this most popular of figures succeeded only in dividing the Republican majority, enabling Democrat Woodrow Wilson to win election with 42 percent of the popular vote. Here, too, the barriers to attaining the nomination, to election, and likely to governance were simply too high.

Campaign Organizations in Candidate-Centered Parties

The golden age of parties endured a long series of changes during the past hundred years that undermined the power of the spoils system to generate the commonality of incentives among office seekers and benefit seekers. As a result, both were led to agree with Vince Lombardi that "winning isn't everything. It is the only thing." Electoral and civil service reforms and related Progressive era "good government" reforms were designed to reduce machine control, that is, to undermine the spoils system. So, too, did the expanding national governmental agenda reduce the relative balance of goods available at the state and local levels in exchange for loyalty to the machine. In particular, entitlements and related devices reduced the machine's ability to give or withhold those goods and services. Each of these changes, and far more, undercut the machine step by step. For office seekers, the growing professionalization of government at all levels made a long-term career increasingly possible and attractive. More than ever before, an entire career in Congress could be carved out. Such goals became more achievable through governmental rather than partisan career paths. As a result, the party counted for less and developing a personal political following counted for more. Thus the bonds that tied both office seekers and benefits seekers to the party and to the paramount goal of winning elections were successively loosened on both sides in a long and gradual process extending over a hundred or more years.

During this period of change, especially the 1940s and early 1950s, scholarly interest in "strong" political parties—typically referred to as more "responsible parties"—was revived. E. E. Schattschneider, in particular, recalled Woodrow Wilson's plea for more responsible parties in the British mold in *Party Government*, and chaired a committee sponsored by the American Political Science Association, whose report "Toward a More Responsible Two-Party System" was published in 1950.[17] By this point, it would seem reasonable to conclude, the bonds that created the golden age of parties had atrophied.

Even so, the basic structure was only weakened, not dead. More importantly, there was no alternative structure for campaigns or for parties generally. Presidential primaries, for example, had existed since 1900. If even the popular Teddy Roosevelt could not win nomination in 1912 by "going over the heads of the party to the people," no one could. In fact, during this period the more votes won in primaries, the worse that candidate did at the convention. Primaries were the route for party outsiders, and outside—and unnominated—they remained. While public campaigns of the whistle-stopping sort became de rigueur for the nominees, mobilization was still done by the state and local party organizations. That is, while there were changes, sometimes important ones, in the conduct of election campaigns, these were harbingers of campaigns to come more than fundamental changes in the organization of either nomination or general election campaigns from the golden age.

A number of advances became progressively regular staples of campaigning. Mass media, an increasingly nationalized and less partisan press and the radio in particular, became more important.

Washington in general and the presidency in particular grew more powerful and became better staffed, providing the raw materials for developing a more personalized campaign organization.

It was not until the 1960s, however, that technology made it possible for a presidential candidate (and soon thereafter seekers of lower offices) to build an effective personal campaign organization. Television, polling, computerized mail lists for fund-raising and get-out-the-vote campaigns, more convenient national travel, and other advances significantly lowered the barriers to entry, making it possible for a candidate to create the requisite far-flung organization all but overnight and to present himself or herself as an individual rather than as a partisan.

Added to these technological advances and the changing incentive structures of ambitious politicians was the decline of parties. An important consequence of this decline was the changing incentives of benefit seekers—the activists who would fill what were and still are labor-intensive roles. These people still depend upon the winners of elections to provide the benefits they seek, but the nature of those benefits has altered fundamentally. The relative importance of patronage has declined significantly, along with the partisan control of channels for allocating distributive benefits. In place of patronage as a source of party activists are what James Q. Wilson has called "amateurs" and Aaron Wildavsky has termed "purists." [18] Their incentives are no longer the distributive benefits of government—the jobs, contracts, and partisan-based policies targeted to areas of party control—but rather more general public policy outcomes. Wildavsky used the term "purist" to describe the legions of conservative activists who filled Goldwater's campaign, but it describes equally well the antiwar activists of McCarthy and McGovern, the "movement" conservatives of Reagan, Robertson, and Buchanan, and many others. That is, the labor is now much more heavily supplied by those seeking governmental policies rather than partisan electoral victories and the patronage and distributive (by party) policies they bring. The patronage-based benefit seeker cared only about whose party made up the government. The contemporary policy-based benefit seeker cares primarily about what government will do. Who makes up the government has become only an instrumental means to that goal. Indeed, the concern of Wildavsky types is that the purist cares too little about winning and too much about ideological purity in its own right. As a result, the party suffers. Umbrella-like parties, providing a home for a wide diversity of interests and concerns and seeking to regulate conflict by giving everyone a half loaf, are unattractive to the purist. A candidate willing to take clear-cut stands on principle and giving every sign of commitment to them is attractive to the purist. Don't choose a whole party with its congeries of interests, says the purist, choose carefully among the particular candidates. Don't just vote for the best candidate and not the party; work for that candidate who best represents your particular interests.[19]

Changing incentives weakened ties to the party for the candidate and activists alike, making loyalty to the party above all an increasingly untenable bond. But it was not until technological changes enabled a candidate to develop his or her own personal campaign organization that the candidate-centered campaign became a reality. The party first lost much of its incentive-induced bond, but it was technological changes that undermined the party's virtual monopoly over the resources (labor and, following that, capital) and made possible a candidate-centered campaign. From a potential activist's perspective, the party as an agent of mobilization has increasingly been replaced by a candidate organization with the resources to provide the selective— and purposeful—incentives to motivate a sufficiently large number of activists (although still a small percentage of the latent pool of potential activists) to staff campaigns. In this context, the reforms that made the presidential nomination process so different in the 1970s from what preceded them were not the product of the self-destruction of the party. Rather, they were the consequence of the changed circumstances. It is nonetheless true—and important—that reforms speeded up the changes, but these driving forces of change in the external environment preceded the reforms. The party leaders did not so much destroy their own role as create a system that reflected and accelerated the changes already in place.

These two views are, of course, simplifications verging on carica-ture of the two eras of parties, but they capture, as do good simplifications, much of the essence of these periods. I refer to the two forms of parties as "parties in control of their candidates and officeholders" and, more contemporarily, "parties in service to their candidates and officeholders." The point is not that the party is irrelevant but that it no longer has the incentive structure and virtual monopoly to be able to *control* its elected officials, as it did when at top strength in the golden era. From time to time, however, it is able to assist them in important ways.

The successful nomination campaigns of John F. Kennedy in 1960 and Barry Goldwater in 1964 illustrate the changes. Kennedy exploited many of the new features of candidate-centered campaigns: overt primary campaigning, polling, creating lists of supporters throughout the nation, the expectation of loyalty to Kennedy, and the various uses of financial resources. He probably invented few if any of these techniques. He did demonstrate their utility, showing the virtues of a positive media image and the special value of television in particular. What was original in his campaign was the successful use of primary victories as key ingredients in his nomination. These were used, however, to demonstrate that his age, inexperience, glamour, and most of all his Catholic faith were not electorally fatal. He used them, as it were, as bargaining chips with party leaders such as Mayor Richard Daley of Chicago and Governor David Lawrence of Pennsylvania to convince them that supporting him would not jeopardize their most cherished goal—winning elections. However, he found too few other leaders, like Daley and Lawrence, who could deliver delegate votes in large blocs. He found the extant party apparatus severely weakened.

Four years later, Goldwater exploited the frustrations of conservatives to out-organize candidates who chose to use traditional party means for seeking nomination. To be sure, it was critical, as it had been for Kennedy, that he win in primaries, defeating that other example of the "good lord," Nelson Rockefeller. Goldwater was also unstoppable because the party, especially its leaders who had been so instrumental in nominating Thomas Dewey and Dwight Eisenhower, could not stop him. With F. Clifton White's support, he used conservatives throughout the nation to defeat the traditional Republican party. Although these were Republicans, they were members of Young Republicans or Young Americans for Freedom, that is, already partially organized but hardly the councils of the Republican party. Out of many, that is, the 1964 Republican contest came down to one "good lord" defeating—and humiliating—another, shaking the established Republican party and altering it forever.

The Changing Nature of Presidential Campaigns and Their Organizations

In the golden age of parties, the disjuncture between the processes of nomination and election meant that there was little nomination organization at all, and except for the nominee's closest advisers, there was very little correspondence between the nomination organization and that of the general election campaign. The nomination consisted primarily of bargaining among the various party chieftains and among them and the candidates' major backers. Bosses were primarily concerned about selecting the candidate likely to run strongest in their region, thereby stimulating the highest possible turnout and level of support at the top of the ticket, furthering machine efforts to win at lower levels. Candidates, of course, cared only about winning nomination and securing the strongest possible support for their general election efforts. Bargains could, in principle, be struck and broken more or less at whim, but the ongoing interaction between machines and victorious candidates made a substantial degree of post-hoc "enforcement" feasible.

Although the general election campaign organization included close advisers of the nominee holding important positions, it was basically an organization of the party, that is, its machines. Not only were these almost the only organizations available, but their effective monopoly was combined with relatively similar incentives. All wanted to win the largest possible victory. Of course, many machines were more secure in their electoral prospects than presidential nominees, especially with strong regional party dominance (Democrats in the South, Republicans in the Northeast) that added up to very close national vote totals in the 1876-1892 period. With victory, the ongoing nature of relations between the president and machines, including members of Congress, afforded many opportunities and incentives to fulfill implicit electoral contracts on both sides. There may have been slippage, but not that much. The incentive and organizational structures, in combination with the normative environment (notably valuing party loyalty above all), yielded candidates aided by the myriad party organizations, but yielded victors relatively constrained by the bonds of party, that is, obliged to act in ways that furthered the party.

From about 1964 on, campaigns changed significantly. First, nominations began to require an organization much more similar to that needed for the fall. In 1960 Kennedy and Humphrey needed public campaigns to make their nomination quests viable. Others did not believe so. By 1964, virtually all serious Republicans had learned the lessons taught by Kennedy's campaign, including the inability to find many bosses able to deliver. The most consequential candidacies were those of Goldwater and Rockefeller, and both well exemplify the new nomination process, but so did others, such as Pennsylvania governor William Scranton in his late bid, rely on a nationwide public campaign. By 1968, only Humphrey campaigned the "old-fashioned way," and like everyone else he felt that he won it the old-fashioned way: he stole it. Reforms simply ensured that all would run the newfangled way.

The most immediate consequence is that a nationwide public campaign for nomination requires a nationwide organization. While local organizations may, from time to time and place to place, be borrowed effectively (be they party organizations or, as utilized by Jesse Jackson and Pat Robertson, other sorts of local organizations), the central organization must be the candidate's own personal organization. Nomination candidates still seek endorsements of state and local party leaders—a strategy that Clinton employed particularly in many southern states in 1992. The value of these endorsements, however, is now more a matter of providing information to the electorate and less of securing the resources and personnel of local organizations.

It is far more efficient to transfer this personal organization from nomination to general election than try to build, from scratch, a wholly new organization. Moreover, there are real benefits for the candidate, notably the advantages of a personal campaign organization that is much more finely tuned to that individual's needs, concerns, and ambitions than any other arrangement. To be sure, primary and general election contests differ in important ways, but the fundamental principle of garnering as much support for the candidate as possible is at least broadly similar.

Much of the modern campaign requires technical expertise on such matters as campaign-related law and the Federal Election Commission bureaucracy, fund raising, polling, mass mailing, media, and advertising. To some extent, this expertise can be purchased through ordinary business means, such as locating an advertising agency or suppliers of campaign paraphernalia. In this regard, the "good lord" with strong prospects and large resources has the advantage. There are two major forces that counteract straight economic, marketplace transactions. The first is specialization, which often follows more or less party lines. Typically, presidential campaigns practically are loss leaders for many campaign consultants, generating a reputation that can prove useful in the much larger number of races for lower office (or even for nonpolitical markets). Market specialization therefore follows partisan lines quite closely, largely because each party specializes in policy appeals and the coalition of voters to whom those appeals (and mailing lists) are especially attractive.

The much more important force for typically following partisan lines concerns the ambitions of the feudatories. Goals are typically not, or at least not exclusively, financial, and this is especially true for the most important experts, whose ambition is often at least partly political. Pollsters, for example, can not only aspire to continuing business by signing on with a winning candidate but can realistically hope to become important advisers in the administration. This is exemplified by Democrat Pat Caddell and Republican Robert Teeter (both of whom have worked with several different presidential candidates, albeit within the same party). Many others can also aspire to achieve influence through business with, or even personal appointment to, the administration. This sort of incentive is even more overt for experts in "movements," which can provide everything from policy expertise to mass mailing lists and related technologies. As described by Schlesinger, these types of benefit seekers want the preferments that association with a successful campaign and office can provide.[20] But they seek policy, not patronage in the traditional sense. Thus the nature of campaign organizations has moved from a feudal structure, with support determined by party and land, to the bastard feudal structure, with support rooted in party and personal ambition.[21]

Incentive structures for the different actors in the golden age of the party, when the machine was working well, were similar. If winning was the only thing, all partisans benefited from more rather than less winning, and all could count votes to see which people were pulling their weight. The stronger the presidential candidate ran, the better the local candidates of the machine would fare. To be sure, the candidate was often subsequently tied by party bonds, but he was in office.

The contemporary campaign is, in a sense, more political in this regard. Surely, all are better off with the candidate winning rather than losing. But the ambitions of today's benefit seekers are far different from those in the golden age. This is especially true for the policy shops and the labor pool, that is, the campaign activists. In the past, these people's desire to win was motivated more heavily by patronage. Today's benefit seekers are much more likely to be motivated by policy. Purists may prefer winning to losing, but they also prefer purity on policy to inconsistency (witness Bush's difficulty in convincing Republican conservatives he is really "one of them," a question that would hardly ever have been asked in the golden age). And the specific issues that concern activists, let alone the policies they support on those issues, are many and varied in a large national campaign organization. The problem, therefore, is that policy-seeking activists often find incentives to diverge from what may be best for their candidate.[22] This is often true even for candidates considered unusually "pure" on policy. At the 1972 national convention, for example, McGovern sought to mollify representatives of the families of POW/MIAs, stating that his plan to end the war would not mean that their husbands and fathers would be forgotten. The Massachusetts delegation, a particularly antiwar slate, immediately called a press conference to announce that they would withhold support unless McGovern reaffirmed his commitment to bringing the troops home within six months. Conservatives whom Reagan mobilized for the 1976 Republican convention went further on platform decisions than even he apparently wanted, leading to public humiliation of Gerald Ford (let alone Henry Kissinger) on the convention floor—even though Ford had the nomination won. All contemporary campaigns, even the most successful, find the candidates struggling to keep in check supporters motivated by policy, perhaps even more strongly than by fealty to the candidate. This concern is accompanied by another that has long plagued campaigns, golden era as well as contemporary: how to keep in check the desire of overenthusiastic supporters to win above all else (as illustrated by the words attributed to Chuck Colson, adviser to Richard Nixon, in 1972, "I'd run over my grandmother" to help elect Nixon).

A national presidential campaign, whether for primary or general election, is today a far more personalized campaign. The nonincumbent candidate has to be able to construct a campaign virtually out of thin air, all but overnight. This is especially true for nomination. Perhaps the single most important constraints on such campaigns are the lack of resources and, quite commonly, the lack of experience in building and running a national campaign. Added to these are the uncertainties about the competition and the electorate's response to the candidates, no matter how thoughtful their strategies. The need to respond to rapidly changing circumstances and to the invariable campaign dynamics only add to these uncertainties. In retrospect, whichever campaign proves successful almost invariably grew at an explosive rate. Brought in to this campaign, therefore, were all sorts of people and resources in short order. Lines of responsibility are thereby strained, and the motivations of those who enter the campaign are varied and often unknown. The problems of compatibility of incentives are, therefore, compounded.

The trick in moving from nomination to general election is how to transform a successful campaign for building support within the party (and against other contestants and forces in that party) into a campaign that faces a different opponent and must appeal to a larger part of the public, who previously could be ignored. This problem is obviously most difficult for candidates whose nomination campaign was contentious and lengthy. I suspect that this is the major reason underlying Martin Wattenberg's observation that since 1972 the party whose nomination campaign lasted longer has always lost the general election.[23] This was the case for McGovern, Ford, Carter, Mondale, and Dukakis, although the same was true for Goldwater in 1964 and Humphrey in 1968. All campaigns of the candidate-centered era, even before the rise of the presidential primary, lost if they fought for nomination longer, so it is not the primaries that mattered. To be sure, intraparty divisiveness always hurts, but so does the inability to shift the organization from winning nomination to winning election. Another aspect of this problem is the necessity to cast aside some tasks and people who were useful in winning nomination but may hinder winning the election.

This describes a campaign built on bastard feudatories, ambitious benefit seekers hoping to realize their goals. It is far from an organization tied to party and thus to land. But if the contemporary campaign organization is not an arm of the party, it is not an army filled with mercenaries because the benefits sought are not (or not only) financial but political, and increasingly based on the currency of general policy rather than distributive policy and patronage. Such political benefits freed of the feudal bonds of party and land are realized by "acquiring a temporary patron. In this loosely-knit and shamelessly competitive society, it [is] the ambition of every thrusting [benefit seeker] to attach himself for as long as it suits him to such as [are] in a position to further his interest." There may be no bonds of party, but political ambitions still require party, ideology, and patrons, if not party, land, and patronage.

The nominee has unusual latitude in designing the campaign organization. There may be responsibilities to reward others, perceived or genuine (a nontrivial part of Ford's early problems), but such obligations are far fewer and less consequential than in the golden age. Indeed, there is little the extant party organizations can command or demand. They can be of service to candidates—presidential contenders have discovered, for example, that state or local party-run registration and voter turnout drives can be of assistance and are exempt from mandated campaign expenditure limits—but they cannot command. The candidate is ultimately responsible for his or her own success, unfettered to a large degree by the bonds of party.

The New Presidential Campaign Organization and Elections

The changing nature of presidential campaign organizations in the candidate-centered era can be illustrated by the differing kinds of tasks and advisers involved in these races. Often the first sign that an individual is seriously considering running for the presidential nomination is the securing of legal services to ensure compliance with federal campaign laws and ballot access in primary elections. Professional advice on the complex structure of delegate selection in caucus states is also an early necessity. If these are not the first indications, it is likely to be acquiring the services of people with expertise in fund raising. Arranging dinners and meetings with potential large contributors remains a staple of nomination politics, but fund raising (and securing more general support) via mass mailing and related technologies is critical.

Another vital early move is hiring a polling organization, whose head often becomes one of the campaign's major advisers. Polling is extensive, and it is common to seek multiple polls in many primary (and some caucus) states. Richard Wirthlin conducted daily tracking polls for Reagan's 1984 general election campaign. Recently, extensive use of focus groups has become an important addition to the task of taking the pulse of the public. These groups are also employed in another increasingly important and complex task that requires professional technical support: advertising. Television consumes the largest part of the advertising budget. Some ads become national news and enter the public consciousness in their own right (Reagan's "It's Morning in America" campaign, Bush's Willie Horton and "revolving door" ads, and those that used the image of Dukakis driving a tank). But there is far more to the ad campaign than television. Whether targeted to specific groups of voters or intended for more general appeals, advertising campaigns employ cable TV, radio, newspapers, and other media. Polling and advertising expertise are often combined with news-media know-how to design strategies ensuring that the story of the day reported will be the one the campaign desires. Speech writers not only craft extended remarks for campaign stops but also write the twenty-second or shorter soundbite to appear, they hope, at the top of the news.[24]

All of this professional expertise (and more, from computer utilization to travel, telemarketing, and the use of satellites) represents additions to presidential nomination and election organizations typifying campaigning in the candidate-centered era. Of course, the more traditional campaign organizational tasks persist, ranging from policy shops, advance arrangers for campaign rallies, dinners, and photo opportunities to the organization of the diverse list of specialized groups supporting the candidate. It is possible, in principle, for a political party, perhaps through its national committee, to provide these services, and to some extent the regular party organizations compete to do so for candidates for lower offices. In reality, at the presidential level (and often at lower levels, too), it is typically the candidate's own organization that secures, and is in large part composed of, these experts.

Many of these consultants, experts, and advisers work exclusively, or almost exclusively, for candidates of one party. Their choice of whom to work for, especially in presidential nomination campaigns, is based on personal ambition as well as party. Indeed, it is precisely this mixture of forces shaping such decisions that define the bastard feudatory system. The personal part of the calculation can be glimpsed in the nomination campaign. If personal ambition attracts the feudatories to the strongest "good lord" in the party's field, such calculations include, but are not limited to, considerations of which candidate appears to be the most likely to succeed.

The extended length of the primary season yields a dynamic campaign often characterized by momentum or similar terms.[25] Unknown or long-shot candidacies may emerge from obscurity to become major, and sometimes victorious, contenders. McGovern in 1972 and Carter in 1976 translated early-season surprises into nominations, while Bush in 1980 went, in his words, from "an asterisk in the polls" (under 1 percent support) to a major rival and eventual running mate of Reagan, and Tsongas in 1992 was originally and falsely written off as an also ran.

The bottom line of such dynamics is the increasing level of popular support and hence committed delegates won. But popular support is only part of the dynamic. With a surprising showing or outright victory in, say, Iowa or New Hampshire, the heretofore long shot attains the credibility to raise money, attract volunteers, and woo political leaders and technical experts to the campaign. After his New Hampshire primary victory in 1992, for example, Tsongas was able to raise more money in one day than he had been able to raise before the win.

Those with skills, resources, and even just time to devote to a presidential campaign are attracted to candidates who show substantial promise of becoming a "good lord" who can make it possible for the feudatories to realize their personal as well as political ambitions. In open nomination contests before the primary season, reporters measure a candidate's prospects by his or her ability to attract the best and brightest of the ambitious to their campaign organization. The candidate with the largest, best, and brightest contingent of experts is often anointed, for good or ill, as the front-runner by the media. And, of course, such a promising "good lord" is not so much attracting the leaders and resources of the national, state, or local party organizations as he or she is building an increasingly strong, effective candidate-centered campaign organization.

Governance in the Era of Candidate-Centered Campaigns

Presidential campaign organizations do not exist in a vacuum. The newer form of candidate-centered organization, for example, has been the product of changes in the external social and political environment. The path to power, however, also shapes the way the victor performs in office. Clearly, the top-level campaign staff often finds its way into the White House, often in political positions but sometimes in critical administrative roles. For example, Bush's chief adviser, James Baker, became secretary of state. This is not new, but the origin of many top campaign staff members *is* new. Such positions are now at least partially filled from the ranks of the bastard feudatories rather than from party organizations.

While personnel is important, perhaps even more important is the transference of modes of operation from the campaign to the White House. Information, ideas, strategies, and tactics are drawn from these feudatories in the campaign, and these critical ingredients are drawn from the same sources in office. The presidential campaign staff relies very little on, and certainly is directed very little from, the party organizations. The party plays as insignificant a command role in the executive branch as it did in the campaign. It may have been of some service to its standard-bearer in the campaign, but exerted little influence. So, too, may the party be of some small service to but exert no influence over, the president. In the golden age of parties, presidential contenders often relied on elected officials of their party, not only for endorsements but also for carrying the major burden of the campaign. Today endorsements by local members of Congress are worth little more to the presidential candidate than the polite thank you and possible honorific appointment as state or local campaign chair. The party platform today, as before, is hammered out in tough bargaining among the various party leaders. But in the golden age, the platform was the initial working agenda of the party in government. Presidents might be able to shape that platform in office, but they could neither dictate nor ignore it. Today the platform reflects the bargaining among the policy-motivated activist elite. It binds no elected official of that party, nor does it constrain their maneuvering room or even define the agenda for any officeholder, president, or member of Congress. The candidate-centered campaign is highly personal. A presidential candidate can make it to office by running against the entrenched leadership in Washington—and even in their own party—as Carter and Reagan did. More generally, presidents come to office without partisan constraints, except insofar as they choose to adopt them.

The organization of presidential campaigns may be better considered the reflection of these trends rather than their cause. But the patterns of decision making in the campaign do have direct impact on patterns of decision making in office, if for no other reason than that the campaign organization becomes a large part of the executive organization. Moreover, the ultimate arbiters of the success of a presidential campaign are the voters in primary and general elections and the media who bring the campaign to the public. So, too, increasingly the ultimate arbiter of presidential influence is the public. As Samuel Kernell analyzed in his aptly titled book, *Going Public,* the president is not only unconstrained by his party and by congressional (often partisan) leadership, there is also less that either group can deliver for the president.[26] Increasingly, presidents have turned to influencing the public to pressure their representatives on preferred policy initiatives.

Their campaign organization was designed to "go public." It worked. In large measure, much of it came to the White House with the successful president. It is only natural that presidents would turn to the White House version of that campaign organization to do what it proved adept at doing in the campaign. The winners in Congress are almost as unfettered by the bonds of party, responsible only to themselves and their constituencies. As a result, these bonds do not tie officeholders together. The policy-making process suffers the fate of 435 + 100 + 1 ambitious, free, and perhaps all too independent officeholders seeking their own goals. Five hundred thirty-six organizations centered on the individuals become 536 individual, autonomous policy makers.

Notes

1. Lewis Chester, Godfrey Hodgson, and Bruce Page, *An American Melodrama: The Presidential Campaign of 1969* (New York: Viking Press, 1969).
2. See, for example, Martin P. Wattenberg, *The Decline of American Political Parties 1952-1988* (Cambridge, Mass.: Harvard University Press, 1990).
3. For a measure of the emergence of the electoral value of incumbency in the 1960s to approximate the value of partisanship in House and Senate elections, see John R. Alford and David W. Brady, "Personal and Partisan Advantage in U.S. Congressional Elections, 1846-1986," in Lawrence C. Dodd and Bruce I. Oppenheimer, eds., *Congress Reconsidered*, 4th ed. (Washington, D.C.: CQ Press, 1989), 153-169.
4. See Gary C. Jacobson, *The Politics of Congressional Elections*, 3rd ed. (New York: HarperCollins, 1992).
5. See also Morris P. Fiorina, *Divided Government* (New York: Macmillan, 1992).
6. David S. Broder, *The Party's Over: The Failure of Politics in America* (New York: Harper & Row, 1972).
7. See Nelson W. Polsby, *Consequences of Party Reform* (New York: Oxford University Press, 1983).
8. See Richard P. McCormick, *The Presidential Game: Origins of American Politics* (New York: Oxford University Press, 1982).
9. For data on these points, see John H. Aldrich and Richard G. Niemi, "The Sixth American Party System: The 1960s Realignment and the Candidate-Centered Parties," Duke University Working Papers in American Politics, no. 107, 1990.
10. Walter Dean Burnham makes this argument in *Critical Elections and the Mainsprings of American Politics* (New York: Norton, 1970).
11. This is modeled after the "textbook Congress" as described by Kenneth A.

Shepsle in "Congressional Institutions and Behavior: The Changing Textbook Congress," in *American Political Institutions and the Problems of Our Times,* ed. John E. Chubb and Paul E. Peterson (Washington, D.C.: Brookings Institution, 1989), 238-266.

12. McCormick, *The Presidential Game.*

13. Richard P. McCormick, *The Second American Party System: Party Formation in the Jacksonian Era* (New York: Norton, 1966).

14. Joseph A. Schlesinger, *Political Parties and the Winning of Office* (Ann Arbor, Mich.: University of Michigan Press, 1991).

15. Theodore Lowi, *The Personal President: Power Invested and Promise Unfilled* (Ithaca, N.Y.: Cornell University Press, 1985).

16. He was also nominated for president by the Populist party, although it refused to nominate his Democratic running mate.

17. E. E. Schattschneider, *Party Government* (New York: Farrar and Rinehart, 1942); the American Political Science Association, "A Report of the Committee on Political Parties," *American Political Science Review,* 44, no. 3, part 2 (September 1950): i-xii, 1-99.

18. James Q. Wilson, *The Amateur Democrat: Club Politics in Three Cities* (Chicago: University of Chicago Press, 1966); Nelson W. Polsby and Aaron Wildavsky, *Presidential Elections: Contemporary Strategies of American Electoral Politics,* 8th ed. (New York: Free Press, 1991).

19. For evidence of the dramatic changes in the balance of "professional" and "purists" in presidential nomination campaigns, see Jeane Kirkpatrick, *The New Presidential Elite: Men and Women in National Politics* (New York: Russell Sage Foundation and Twentieth Century Fund, 1976).

20. Schlesinger, *Political Parties,* 1991.

21. It is important to emphasize that it is not just personal ambition that defines the analogy. If that were so, the analogy would be to mercenaries rather than to bastard feudalism, and if it were so, the best and brightest of the ambitious would, over time, gravitate to that candidate offering the strongest incentives for realizing personal ambition, the candidate with the best chances of winning the election, regardless of party or any other aspects of the "feudalism" in bastard feudalism.

22. This problem is far worse for a party, which includes many different candidates, than for a single candidate.

23. Martin P. Wattenberg, *The Rise of Candidate-Centered Politics: Presidential Elections of the 1980s* (Cambridge, Mass.: Harvard University Press, 1991).

24. The original film of Dukakis driving the tank, which became the well-known Bush advertisement, was devised by Dukakis's campaign organization as the video to accompany their message-of-the-day on defense policy targeted to be the lead story on the candidate on national TV news.

25. See John H. Aldrich, *Before the Convention: Strategies and Choices in Presidential Nomination Campaigns* (Chicago: University of Chicago Press, 1980); and Larry M. Bartels, *Presidential Primaries and the Dynamics of Public Choice* (Princeton, N.J.: Princeton University Press,

1988).
26. Samuel Kernell, *Going Public: New Strategies of Presidential Leadership*, 2nd ed. (Washington, D.C.: CQ Press, 1993).

3. THE PERSUASIVE ART IN POLITICS: THE ROLE OF PAID ADVERTISING IN PRESIDENTIAL CAMPAIGNS

F. Christopher Arterton

The consensus among presidential campaign professionals is that, in the words of political pollster and consultant Pat Caddell, "the message is everything." A campaign must first decide on a message for the candidate, the content of which is dependent upon the audience.

Christopher Arterton argues that candidates distinguish between two types of voting publics: committed voters, who form a loyal "base" for a particular candidate; and uncommitted voters, for whom the candidates compete. The biggest potential problem in designing a message is that the most attractive message to uncommitted voters may alienate the loyal supporters. Appeals to uncommitted voters therefore risk moving a candidate too far from his or her original base of support.

Candidates have two strategies open to them: They can redefine their own message, or they can attempt to redefine those of their opponents. If they try to redefine themselves, they are likely to redefine their base of support as well, without necessarily expanding it. Moreover, the more they attempt to reshape their message, the more they run the risk of leaving all voters in doubt about what their message is. As pollster and political strategist Greg Schneiders noted, "it is easier to drive negative numbers on the opponent than it is to build positive numbers for yourself." Thus, candidates find attacking their opponents and their opponents' messages, through negative advertising, safer than trying to expand their own appeal. The dangers, however, are twofold: (1) Voters do not like negative advertising; and (2) Voters generally feel alienated by political campaigns, which now are dominated by negative messages.

Arterton argues, as a result, that support for candidates in the electorate grows increasingly shallow as candidates undermine the support of their most loyal backers. Furthermore, he says campaigns and elections no longer serve to align voter preferences on any particular dimension. Instead, campaigns contribute to a continuing process of "dealignment" in the electorate, as voters fail to see good reasons for adhering to any candidate or party. —Ed.

Asserting that presidential campaigns need to communicate is akin to observing that water is wet; campaigns are little more than vehicles for communication. In this, they may be different from party organizations to which we might legitimately impart a variety of other political functions such as candidate recruitment, coalitional bargaining, or interest aggregation. Defining audiences, designing messages, conveying information, imparting perspectives, and gauging the effects of its actions upon voters are, in fact, the essence of a campaign, the reasons why a candidate needs an organization in the first place. How effectively the organization achieves these tasks may well have a pivotal impact on the outcome of the election.

Though campaigns may be devices for communication, pure and simple, we should not assume that the process of political communication will be all that simple or our understanding of that process all that pure. To comprehend the communication needs of politicians, we might begin by considering the array of overlapping audiences with whom simultaneous and more or less continuous contact must be maintained. Candidates for public office assemble campaign organizations in order to communicate with

- campaign workers in order to coordinate activity and sustain morale
- journalists both as intermediaries to the public and as key players in establishing the perceptual environment within which the campaign must compete
- party leaders and other public officials in order to obtain endorsements, support, and help for their campaign
- potential contributors, large and small, in an effort to raise the funds needed to contest the election
- interest-group leaders and members to describe policy positions and solicit support and help in campaigning
- finite segments of the electorate (defined according to demographic characteristics, geographical boundaries, or political beliefs) in order to influence their voting behavior
- strong party identifiers in an effort to encourage a high rate of turnout
- all potential voters in order to win their support and maximize the favorable vote on election day

Since this essay focuses on television advertising by presidential candidates, only a small part of these total campaign communications is discussed here. In the process, I wish to explore as well differences in perspectives of political scientists who seek to understand presidential elections and of political managers who seek to affect the outcomes of those contests. The gulf is wide. By way of illustration, a recent textbook on persuasion theory contains a lengthy section devoted to "Compliance-Gaining Message Production," a label that most political consultants would not even recognize as a description of their business.[1]

The differences in perspective run deeper than the jargon employed by each profession. One major disagreement, for example, needs to be faced at the outset because it frames the overall argument of this essay. Do presidential campaigns matter? An important field of political science employs statistical modeling to relate voting in presidential campaigns to long-term social and economic trends, such as the growth in real per capita income.[2] According to Steven Rosenstone, for example, the Bush campaign against Dukakis "was really smart, and the big negative campaign . . . was a wizard show, [but] any competent campaign that played up the state of the economy would have gotten basically the same result."[3]

A literal interpretation of these econometric models would suggest that they demonstrate conclusively that the campaign itself is of no importance. Naturally, political managers would vigorously dispute this assertion. Even while acknowledging the pivotal importance of the economy in framing the context of the general election vote, they assert that events in the campaign and the messages they produce in advertising will shape the choices made by critical sectors of the electorate. Historical events—short-term factors in the jargon of these models—can always intervene to shape electoral outcomes.

Even where longer-term trends may establish the general framework of election choice, a great deal goes on within that framework as the competitors focus their energies on the marginal group of voters who will tip the election. The conduct of candidates and their advisers during the campaign becomes, moreover, an important component of the rich folklore that develops to explain what happened and thus takes on a reality of its own. In replaying the 1988 elections, for example, accounts are very likely to cite George Bush's attacks on Dukakis for his refusal to require the pledge of allegiance, his support for the American Civil Liberties Union, and the Massachusetts prison furlough program. Well into his first term, Bush was still paying a price among some segments of the electorate for his conduct during the 1988 campaign. Even as the 1992 campaign opened, research conducted by one strategist showed the effects of an electorate still reacting to the last campaign by demanding more specifics from candidates and a more serious treatment of issues than mere sloganeering.[4] At least in the popular impressions that often define political action, the long-term factors are very far down the list in our understanding of what occurred in 1988.

In analyzing campaign communications, we must address more than the different audiences that candidates need to reach. We will also consider the messages that campaigners endeavor to transmit and the media they use to communicate. These three components of the communications process—audience, medium, and message—depend, in important respects, upon such factors as (1) the level of office sought, (2) the nature of the election (primary, general, or runoff), (3) the strategic circumstances of the candidate and the relevant opponent(s), (4) perceived strengths and weaknesses in the character of the candidate, (5) the level of financial resources available to the candidate, (6) public thinking and opinions, and (7) the candidate's commitment to a range of ideological positions and public policy preferences.

Complexity of the magnitude created by this wide range of factors exposes enormous gaps in our understanding of the general field of political communications. In this area, theory is unusually weak and detailed research disappointingly sparse, perhaps because a very large number of variables must be considered while the number of observation points remains quite small. Taking into account variations in the pivotal audiences, substantive messages, alternate media, and strategic circumstances of the campaign, we can think of an infinite number of possible relationships to be examined by reference to a finite number of campaigns. If one picks 1960 as an arbitrary starting point for modern presidential campaigns, for example, the total number of general election candidates up to the 1988 campaign is only eighteen (including the third-party efforts of George Wallace in 1968 and John Anderson in 1980). Over this same period, the number of candidates seriously contesting their party's nomination to the point of running advertisements on television was certainly less than seventy-five, far short of the richness of data necessary to generate quantitative insights into a complex process.[5]

At the same time, so much has changed the face of political broadcasting over this period of study that we might well question whether 1980 rather than 1960 would provide a more suitable starting point for our analysis. As videotape has replaced film, as satellites have made possible live broadcasting from remote locations and teleconferencing to a large scattered audience, as cable networks have joined broadcast licensees, as network audiences have declined, as politicians have learned that they can get away with so-called attack advertising, the nature of presidential campaign advertisements has responded so significantly as to undercut the validity of observations based on earlier campaign years. Despite these difficulties, the topic is important enough to demand our attention. The mere fact that presidential candidates almost without exception pour a high proportion of their campaign treasuries into television impels us to undertake as detailed an examination as possible. According to Herb Alexander and Monica Bauer, of the $46.1 million that the two 1988 presidential campaigns were awarded from the public treasury, the largest item in both budgets was media purchases: Dukakis spent $23.3 million on media (or 50.5 percent); Bush $31.4 million (or 68.1 percent).[6]

In this essay I will address the issues raised by television advertising under four broad topics: (1) the questions that surround the strategic use of campaign ads; (2) the nature of the advertising content; (3) the problems and policy issues that emerge from a discussion of campaign advertising; and (4) possible alternatives that would allow effective political communications while minimizing the unfortunate characteristics of current practices. In effect, the first two topics will require us to lay bare the logic of contemporary campaigners; the latter two will cause us to reflect upon the political and policy consequences of how they conduct their business.

The Tactical Use of Political Ads

A basic aphorism governing any form of communication counsels the speaker to start by defining the audience. Knowing with whom to communicate will move one a long way toward knowing the message to be delivered. For the purposes of this essay, such advice implies an opening discussion of the strategic goals of presidential campaigns and their tactical use of advertising, even though questions of content may be a good deal more interesting.

As a potential audience, the vast size of the American electorate constitutes a daunting task. Seen through the eyes of political consultants, the first step in defining a campaign strategy lies in reducing the overall electorate of 185 million to a target audience for campaign communications by sorting out likely or probable voters from the population of adult citizens.[7] The various techniques for making this cut have become so routine that campaigners rarely question the implicit logic. As they see it, the campaign should be pointed directly at those who will probably vote anyway. On the local level, most campaigns still rely on age-old procedures of defining their universe through lists of registered voters, which they upgrade by the laborious process of culling the public records of actual turnout in past elections. Consultants on the presidential level are more apt to place their confidence in random sample surveys in which the first questions are designed to screen out respondents who probably will not vote. Thus, as a decision tool of campaign strategy, political polls normally generate messages circumscribed by the interests and predispositions of likely voters.[8] In either case, the campaign becomes directed toward a limited sphere of the potential electorate.

In terms of contemporary American politics, this may well be a fateful decision, for, rather than stimulating turnout, the logic of modern campaigners directs all of their activities primarily toward influencing the behavior of those who will get themselves to the polls. While we cannot lay all of the blame for declining rates of participation in American elections at the feet of campaign strategists, the predominant form of political organizations at work in electoral processes today does very little to encourage turnout.

We cannot completely ignore voter registration drives. Under the constraints imposed by state laws, both political parties devote energy and resources to increasing the roster of voters in advance of the fall campaign. Nevertheless, in the mind-set of contemporary campaigns, these efforts are almost always seen as marginal, based on campaigners' general perception that direct face-to-face contact with voters has limited value. Questioned as to the impact that "organization" might have on the final election results, campaigners almost uniformly estimate that no more than 5 percent of the vote can be influenced by this aspect of the campaign.[9]

Having narrowed their ken to the pool of likely voters, campaigners become primarily absorbed in the task of reaching a particular segment of that audience—the undecided voter. In designing strategy, political managers categorize the electorate: *our* supporters whom we cannot lose, *their* supporters whom we cannot win, and the middle group of persuadable voters whom we must attract. Logic directs them to three simple rules: reinforce our base vote, ignore their solid vote, and concentrate energies on the undecided voters.

Logic is one thing; in practice the mathematics of base and uncommitted voters may fluctuate dramatically, depending upon such factors as the stage in the election cycle, the level of office involved, and the mix of candidates running. Nevertheless, in general elections presidential campaigners believe that they can specify these segments with some precision. For example, Edward Rollins, who served as campaign director for Reagan-Bush in 1984, observed, "In most years, we've got 41 percent, and they've got 41 percent, and the race boils down to the 18 percent in the middle." [10] His Democratic counterpart that year, Robert Beckel, who managed the campaign for Walter Mondale and Geraldine Ferraro, basically agreed with this assessment, lamenting only that he had lost most of the pivotal voters to Rollins.[11]

Can we find any evidence to validate this model? For one thing,

Table 3-1   Timing of Voting Decision for Presidential Candidates, 1952-1988

	Percentage of voting decisions for presidential candidate made		
Election year	Before conventions	During conventions	After conventions
1952	36	32	32
1956	60	19	22
1960	31	31	38
1964	41	25	34
1968	35	24	41
1972	45	18	37
1976	34	21	46
1980	40	18	41
1984	52	18	31
1988	31	28	39

SOURCE: Warren E. Miller and Santa A. Traugott, *American National Election Studies Data Sourcebook, 1952-1986* (Cambridge, Mass.: Harvard University Press, 1989), 319; and the 1988 Election Survey.

NOTE: One caution about accepting these data too literally is warranted: these are after-the-fact recollections. Voters were asked in the postelection survey when they had made up their minds. Their recollections may be faulty.

recent history imparts at least an impression of confirmation: the popular vote in presidential campaigns since 1960 has normally fallen within these parameters.[12] Even when one party nominates a weak candidate—as did the Republicans in 1964 and the Democrats in 1972—the division of votes has not fallen substantially outside this "sixty-forty" split. Another bit of support can be gleaned from voting studies conducted by political scientists: even in the face of the weakening influence of party identification, these surveys demonstrate that the behavior of a substantial portion of voters can still be predicted by the "normal party vote." [13]

The candidacy of an incumbent president is likely to introduce even more solidity to the base votes for each side, as can be seen in the first column of Table 3-1. When an occupant of the White House is contesting the election, on average about 12 percent more voters report having decided whom to support before the nominating conventions.[14] Incumbents become well known, so that by the time they run for reelection, after years in office, they tend to be either highly or poorly regarded by most voters. If retrospective assessments characterize voting in most presidential contests, these assessments are likely to be even more dispositive when an incumbent stands for reelection.[15]

Finally, we might postulate that as the strength of party loyalties has weakened among voters over this time period, the number of probable voters potentially up for grabs is likely to have increased. Two observations based on this survey data support this proposition. First, as can be seen from the third column in Table 3-1, with the exception of President Reagan's strong early support in 1984, in general the percentage of voters who report choosing their candidate during the fall campaign has increased since the 1950s. Moreover, when the answers to this question are examined in terms of strength of party identification, we discover that much larger numbers of weak identifiers report deciding late in the election cycle.[16]

Though campaigners may think they can estimate the approximate size of the base and swing vote, the process of defining target audiences is not all that simple. The dynamics of voter choice are a good deal more fluid than this description implies. For one thing, citizens may make interrelated but conceptually distinct decisions as to whether to vote at all and, if so, which candidate to support. It seems reasonable to assume that, in contrast to those who feel strongly about a given candidacy, undecided voters are less likely to turn up on election day. As a result, the pool of probable voters keeps shifting, as those without strong attachments toward any of the candidates waiver in their commitment to make it to the polls at all.[17] Furthermore, voters who are initially committed to one candidate but have some reservations may, in fact, shift into the undecided category as the campaign unfolds and may subsequently reaffirm their initial decision to vote for that candidate. A few voters may even change allegiance from one candidate to the other as events cast the race in a new light.

In short, though the concepts of base and swing voters may have a surface appeal to campaigners in developing their persuasive strategies, neither of these two groups is rigid enough to provide a definitive guide to the audiences that campaigns need to reach. On the contrary, there is good reason to be cautious in accepting this model, particularly as a mechanical formula that can easily be followed.[18]

Nevertheless, the notion that campaigning ought to be directed toward the small segment of undecided voters is fundamental to the creation of an advertising strategy. If only about half of the eligible citizens constitute the pool of likely voters, the television advertising of presidential campaigns will have as a target audience roughly 10 percent of the nation's adult population. Toward this market they will direct upward of 60 percent of their available resources.

Having defined this group, in short order campaigners want to know the demography of undecided voters. Characteristics that can be linked to geography are particularly useful: undecided voters are more likely to be found in suburbs or in particular kinds of communities or—

on the presidential level—in states with a large number of electoral votes. So-called geodemographic targeting tries to relate political attitudes and lifestyle variables to residential patterns.[19] In other words, survey data to pinpoint the audience of undecideds is useful both in framing a persuasive message to be delivered and in selecting among the various channels of communication for reaching those voters.

Once they have a precise idea of whom they wish to reach, presidential campaigns confront a problem. On the basis of efficiency, paid broadcast television commercials are the preferred vehicle for delivering the campaign's appeal.[20] But the reach of broadcast television is so extensive that the idea of tailoring TV ads to 10 percent of the adult population is not very practical. Their chosen instrument for communication turns out to be too broad-gauged to isolate these groups very effectively. Broadcast television reaches massive audiences that are comparatively undifferentiated in terms of demographic groups that might be useful to campaigns.[21] Instead, television is especially valuable in communicating with the entire pool of probable voters.

Accordingly, as a tactical tool, television presents some very real challenges to campaigners who would implement the logic of appealing simultaneously to base and undecided voters. To the extent that each group has a distinctive constellation of opinions and values that must be appealed to, campaigners risk alienating one element while communicating with the other. If the messages seek to reassure base voters, they risk boring or alienating the undecided. Conversely, messages designed to bring the undecided into the fold may do very little to galvanize turnout among base voters.

There are two solutions to this dilemma. First, campaigners can use broad, unifying themes or symbols that will appeal to all and offend none. Second, they can encompass both audiences by attacking their opponents. As we shall see, television advertising is roundly criticized for both strategies.

On the level of presidential contests, geography provides some assistance. When demographic differences in a target audience line up with the geography of media markets, a campaign can use television more efficiently. In fact, the larger the geographical scope of a campaign, the more likely this is to be true. Presidential campaigns can, for example, direct messages toward undecided voters who are clustered in media markets located in states with pivotal electoral votes.

It is also true that political consultants can tailor their media buys to some extent according to the demographics of audiences. The audience for sports programs tends to have a higher proportion of males; daytime programming draws a disproportionate number of female viewers; those who watch news programs are much more likely to be voters, and so forth.[22] But these differences are not absolute. The resulting audiences are not composed of homogeneous clusters drawn from given social groups; rather they are more appropriately pictured as statistically significant deviations from national norms.

For presidential campaigns, the cost of reaching voters by television poses another dilemma. Purchasing a national audience by placing an ad in prime-time network programming constitutes the greatest efficiency in terms of cost per thousand voters. Yet the local market, the so-called spot market, offers efficiencies both in terms of the mathematics of the electoral college and the greater variation in the audience demographics of non-prime-time programming.[23] Accordingly, campaigns frequently adopt a two-tier strategy for the general election in which the national buy provides an umbrella by conveying the most unifying messages to the most general audience. Meanwhile, buying in the spot market allows the campaigns to hone in on the pivotal undecided voters with messages specifically tailored to their concerns.

In reaching voters through television, political advertisers engage in overkill. They want to reach voters repeatedly. Merely to maintain visibility in a race, they will buy around 200 points a week.[24] To "drive the numbers," campaigns buy upward of 1200 rating points in a week; instances have been reported in which the campaigns of both parties neared the saturation point, buying over 5000 points the week before the election.[25] While academics may doubt the effectiveness of political commercials, time and again campaign consultants report measuring a gain in visibility and popularity for candidates who can afford to advertise at between 600 and 1200 rating points a week.

Presidential campaigns turn to broadcast television commercials within a larger context. Though predominant, TV ads are not the only means available for reaching target audiences. Moreover, some of the other instruments can be more finely tuned to reach undecided voters. The direct-mail industry, to cite a prominent example, has achieved an elaborate capacity to designate the characteristics of individual recipients. If, for example, campaign consultants learn through polling that worries about the deterioration of the environment are highly salient among undecided voters, they can choose several approaches to convey their message to this audience. In addition to cutting a TV ad addressing these concerns, they may work through existing environmental groups to contact their membership. They can mail or call all households in neighborhoods where residents have the characteristics of those likely to be concerned about the environment (geodemographic targeting). They can also purchase lists of consumers in key states who have checked environmental concerns on the lifestyle indicators that accompany most warranty registration cards.

Even though mail is more efficient in reaching voter segments, it is generally considered less effective in influencing voters. It is nonetheless but one example of an expanding array of media that campaigns have at their disposal. Future presidential candidates will undoubtedly make greater use of the channels that are often referred to as the "new media." [26] Of particular importance to the topic of this essay, we should anticipate a rapid proliferation of messages distributed on videotape cassettes in the coming election cycles. Such products give politicians three advantages: the ability to target as effectively as mail, combined with the power of the visual medium, magnified by complete control over the design of the message, which does not have to pass through any filter.[27] Many of these videotapes will be products of the campaigns; others will be pseudo-documentaries produced by groups and organizations that are either totally or formally independent of the campaigns.[28]

Thus far we have neglected the fact that election contests involve more than one speaker and an audience. Campaigns also involve opposition, a fact rarely taken into account in the academic literature on persuasive communications. In the tactical world of the political practitioner, advertising rarely occurs without direct opposition. As Kathy Ardleigh advised, "The first rule of political advertising is never let your opponent have a free shot at you; never let them be on the air without a response from you." [29]

Recent technological advances have greatly facilitated the tactical maneuvering of campaigners. The arrival of videotape and minicams has so sped up the production time in advertising that campaigns can now respond to each other rapidly. Not so long ago, campaign ad programs were rather like set-piece warfare: candidates and their advisers would plan their ads well in advance, film and edit them, and then run them in the last three weeks.[30] No longer. Now they change their ads constantly as events in the campaign unfold. Fielding a response to an opponent's ad within twenty-four hours is notable but by no means exceptional. As a consequence, the back-and-forth banter between candidates has expanded from speeches and interviews into their advertising as well.

The resulting competing pulls on voters render the effectiveness of ads very difficult to assess. Moreover, while the conflict between candidates can be quite visible in the content of their ads, that is only part of the story. A considerable amount of jockeying and combat goes on behind the scenes, out of the public eye. Federal laws regulating the candidates' access to the airwaves have the effect of allowing each campaign to monitor its opponent's time purchases. As a result, campaigns have organized elaborate feints to throw off opponents, such as reserving time very early in the election cycle in the hope that one's opponent will match the buy. Just before the ads are scheduled to run, the buy is canceled. If all goes well, the opponents will not react fast enough and will spend a large chunk of their money before the race really begins. To cite other examples from the 1988 primary campaign, competitors reserved TV time in certain states in order to make their opponents think they planned to contest that state seriously, only to cancel the buy just before the primary. When it came to the South Carolina primary, the Dole campaign used the same ads they had been running in New Hampshire but listed them in their ad buys under new titles, one indicating that the ads were an attack on South Carolinian Lee Atwater, the Bush campaign manager. The Bush campaign reportedly spent a day getting ready to respond to an attack that never came.[31]

The tactical combat between contestants can, of course, spill over into the content of the ads themselves. The 1988 Bush commercial on pollution in Boston Harbor may well be a case in point. As Pat Caddell, who conducted polls for Jimmy Carter, observed, "I think the Boston Harbor spot was also directed, whether consciously or not, at screwing up Dukakis's head as well as the press, by showing that they could come [to Boston and] do something really interesting, and throw Dukakis on the defensive. It may have been as effective in [doing] that as it [was] in moving a lot of votes." As these few examples illustrate, presidential campaigns use their advertising for tactical advantages that go well beyond their skill in isolating and communicating with appropriate audiences of voters.

Content and Strategy in Political Advertising

The insider gamesmanship surrounding the tactical use of political ads is not, of course, the most visible aspect of the sport. Observations about the persuasive campaigns of presidential contenders most frequently concern the content of the messages directed toward voters. Over recent election cycles, paid advertising in politics has become the focal point of ever intensifying critiques, reaching a crescendo after the 1988 campaign, which many viewed as negative and addressed to trivial issues. This mounting concern impels us to ask, "What do politicians have to say to the American people in their ads?"[32]

Questions of content are very difficult to address empirically because they inevitably force us into a subjective world, evaluating messages sent by the candidates and communications received by the voters. We cannot rely upon the stated goals and objectives advanced by the candidates and their advisers, who cannot be trusted to be totally candid, particularly while the campaign is in progress. Nevertheless, a topic of such importance cannot be ignored.

Media consultants employ a jargon to delineate frequently employed types of ads. Rather surprisingly, this language is neither very rich nor very precise. The terms used to describe the content of their ads seem better designed to indicate rough categories rather than to communicate substantial knowledge about their work. Categories rarely extend much deeper than the obligatory "bio spots" used to introduce viewers to a candidate by presenting past experiences in a favorable light, "comparison spots" in which the records or proposals of two opponents are starkly contrasted, "person-in-the-street ads" in which supposedly average voters talk about the candidates, or "attack ads" in which one aspect of the opponent's past or program is singled out for condemnation.

Here academia can provide practitioners with a germane and useful assist through descriptive studies that offer several complex schema for thinking about different types of ads.[33] Johnson-Cartee and Copeland, both of whom double as political consultants, provide a highly elaborate categorization of political ads.[34] For example, noting that ads frequently attempt to blur the boundaries between paid spots and news programming, they describe eleven different presentations such as person-in-the-street interviews, news interview shows, the visual documentary, and so on.[35] Until such a vocabulary percolates down to practitioners, critics, and observers, however, the language used to describe the techniques employed in TV spots will be inadequate.

Agreeing with Daniel O'Keefe that very little is known about how media communicators go about their work, in an effort to illuminate their logic, we might profitably examine the procedures followed by consultants in producing ads.[36] As in all walks of life, there is an ideal, and then actual practice. Given the resources usually available to presidential campaigns, if the ideal model is followed anywhere, it is likely to be at the presidential level.[37] From its genesis in attitudinal research, political advertising strongly resembles product advertising, although the instrumental focus and hurried pace of campaigners can generate some very different approaches. Using a combination of random sample surveys to measure the distribution of political sentiments and focus groups to examine the complex details of citizen thinking, consultants sometimes deliberately violate some accepted rules of objectifiable academic research. For example, pollsters often use a loaded hypothetical question to test the impact of a persuasive message: "If you knew that the candidate had used political influence to get out of the draft, would that affect your vote?" Similarly, focus group leaders will often solicit the opinions of participants and then challenge them to respond to one-sided information about the various contenders.

Crude though they may be, these pretest-stimulus-posttest research designs, which must be transparent to the respondents, have guided the creative process behind numerous political ads.

Despite diminished standards of reliability and validity in their investigations, a good deal of research lies behind the creative process of packaging political messages. Most campaigners recognize, for example, that they themselves live in a rarefied world of extreme attention to political debates. Therefore the communications they produce may not always translate into messages that are meaningful to voters or effective in presenting the intended argument. To compensate, in many cases the finished ads will be shown to focus groups before they are ever put on the air. In a few cases involving presidential campaigns, ads have been test marketed before being used nationally.[38] Finally, so-called tracking polls—the latest weapon in campaign survey research—have been developed to monitor the impact on voter thinking of campaign events of all kinds, including ad campaigns.[39]

As might be expected, while political managers are anxious to know what works in influencing voting decisions, they are not very interested in *how* these communications work. They rarely take the time to contemplate systematic theories of communication or persuasion. Operationally, persuasion theory remains fundamentally a derivative of cognitive psychology, rooted in the processing functions of the individual receiving the communication.[40] Meanwhile, politicians concentrate on the message and the audience as a collectivity. To be sure, many creators of paid advertisements would admit that their success could be improved by knowing how individual members of the audience process their messages, but they are too busy to treat the citizen as anything more than a "black box." They concentrate on the effects produced by their messages, not on explanations of how their results are achieved.

Accordingly, consultants are noticeably vague in describing how their ads actually work, and very little theorizing—either explicit or implicit—undergirds the advertising of presidential campaigns. They make one thing very clear, however: television ads primarily work their magic visually rather than aurally. On this point, the creators and critics of advertising agree. Douglas Bailey, Gerald Ford's media adviser, observed:

> I would stress that television is a visual medium, and, as much as the words—the information—may be important, it is the visuals that you want to stay in people's mind. . . . You're talking symbols, symbols, symbols. . . . The first question on the check list is, "What do the visuals really say?" Turn the words off, turn the sound off, what do the pictures say, because that's what is going to be remembered in people's mind.

Media critic Mark Crispin Miller lamented that "these images just kind of flash onto your consciousness. That's really all it takes, you see. Because you're not meant to really scrutinize these things. . . . Once you do scrutinize them, then you can see just how destructive and inhumane this pitch has finally become." [41]

While media specialists may agree that television ads work primarily as visual communications, they disagree on the cognitive mechanisms through which ads achieve an impact on voters. Much simplified, we postulate that ads might achieve an impact on voters by providing information that alters their cognitions, by reshaping attitudes toward candidates and their programs, by affecting the prominence that different issues or questions have for voters, or by linking the candidate in question to preexisting deeply held values. The first approach is educative; the second is persuasive; the third focuses voter attention; and the last assumes that ads work by serving a cuing function for voter predispositions.

Even though as a practical matter most campaign ads contain each of these elements, individual consultants in political advertising generally emphasize one of them in their work. Most consultants, however, reject the first approach wholeheartedly. Unquestioned dogma within this community counsels that campaigns are no time to start educating voters. As Douglas Bailey has observed, "One thing you're not doing is you're not educating voters to a wholly new idea or concept. You don't have enough time in a campaign to do that, and you certainly don't have enough time in thirty seconds." Take any complex issue, such as health care. Candidates may bemoan the deteriorating access to insurance; they may play upon voter fears of rising costs; they may enunciate detailed programs in the attempt to convince voters and journalists that they know what should be done; they may contrive appearances that heighten the issue and identify their candidacies with an effort to ameliorate the problems, but they rarely seek to educate voters in the nuances of the problem or the details of alternative solutions.

Strictly speaking, political commercials do impart information. On an elementary level, they announce to voters that somebody is running for office; more elaborately, they may communicate rudimentary details about, for example, the opponent's record. [42] Here again Doug Bailey provides an illuminating glimpse of the logic employed by media advisers:

> The one key of almost any ad is linkage. That is, you've got to tie it together with something that is already in the viewers' mind. You've got to make a link somewhere so that the ad is seen as

sitting in some context that the viewer can understand and is likely
to believe. Secondly, once having made that linkage, then you can
add a little bit to the information that they [already] have, in order
to make more of a point.

He went on to discuss this process, using the "person-in-the-street" ads
he created for Gerald Ford in 1976:

[T]hey were linkage; they played to a concern that voters already
had. . . . In Atlanta, we interviewed over 500 people on camera. We
measured very carefully what they had to say and the words they
used, against what the polling said people already believed. . . . [If]
their statement didn't seem to link up with the thinking that people
already had, we eliminated them.

Though such comments may cause us to argue over the accuracy
and meaningfulness of the communicated information, clearly a major
component of any ad campaign consists of altering voter cognitions by
transferring information. The difference lies in how one defines
education. If that implies communicating extensive amounts of information that will substantially deepen the knowledge base of many
voters, then campaigns will always fail the test. The time available to
them, the resources they can muster, and the essence of the medium are
simply inadequate to the task.

For the same reasons, media managers are uncomfortable accepting the persuasive approach too readily. They do not feel that, in the
midst of a presidential campaign, they have the capacity to reshape
substantially voters' beliefs or attitudes. Campaigners would acknowledge that, over time, public debate may influence the distribution of
preferences on such matters, but they insist to their clients that
campaigns should not attempt to persuade voters to change their views.
While shifts may occur in the course of a campaign, that is incidental to
their approach.

This is not to argue that presidential campaigns steer clear of
public policy issues. On the contrary, media advisers are constantly in
search of issues in which the majority of targeted voters—be they the
entire electorate, the candidate's supporters, or the undecideds—agree
with the stated position of their client and disagree with their opponent.
These are the issues that need to be pushed to the forefront of the
contest. In other words, much political advertising follows the third
approach described above: ads are intended to set the agenda for public
discussion and consideration, not to influence attitudes about public
issues. From this perspective, the whole campaign—including both
paid advertising and public relations efforts designed to influence news
reporting—is an effort to raise the visibility of advantageous issues. As

media specialists see it, campaigns achieve success by "driving the issues," "dominating the definition of decision," or "setting the agenda." "Controlling the debate"—another phrase they use to describe this process—refers less to winning arguments than it does to confining the public discussion to a range of issues that favors their candidate.

This effort to advance the salience of some issues over others could be labeled an agenda-setting approach, establishing a connection between the campaigner's strategy and the conclusions among academics over the electoral effects of news reporting. An abundant literature emphasizes that news reporting rarely has simple persuasive effects on voters' thinking but can have pronounced influence on what voters think *about*.[43] In response to the inevitable criticism that their work inadvertently determines elections through their choice of issues to cover, journalists argue that they rarely raise issues in a vacuum but rather respond to events in the world. That many campaigners use television ads to focus attention on some issues while obscuring others provides some support for the journalists' point of view. Campaigners wish to influence the agenda of public thinking both directly through their ads and indirectly through news reporting either about the ads themselves or about the issues being raised by candidates in their ads.

Media consultants who embrace the salience approach may differ over the details of how this goal is best achieved. One dispute concerns the nature of the issues that are most profitably raised. For example, many believe that personal aspects of the candidate are much more important to voters than issue positions. To criticism that they advance images rather than address public policy, consultants often cite the adage that, since policy issues of the day change rapidly, the candidate's character is a better guide to future performance.

Some consultants believe that campaign content should be designed so as to frame a question in voters' minds: "What is this election all about? What is being decided by my vote?" Advertising serves not to communicate information but to raise questions. In practice, reducing the campaign's persuasive message to a single question may simply be a strategic device to focus clients on a limited agenda of issues. An important subtext of this approach to advertising counsels that campaigns should stress only a very limited number of issues to avoid public confusion. Focusing the client on *the* question that voters should be thinking as they approach the voting booth may be useful in reaching a well-defined campaign strategy.

Finally, though many consultants agree that advertising works by raising the salience of issues, they disagree on whether ads should advance support for the client or erode support for the opponent. The decision to use negative advertisements results largely from strategic calculations of the individual circumstances of the presidential candidate. Nevertheless, the recent rise in the popularity of these ads indicates something else has occurred.[44] Political consultants have learned that, within limits, the public will tolerate negative commercials, especially if they address political issues rather than the personal characteristics of the opponent.[45] In addition, negative ads often prove to be very effective.[46] As noted by Doug Bailey, "A vote for president requires all kinds of positive images. One good point isn't enough. . . . One bad point against is enough to rule him out. Consequently, it's much easier to use my thirty seconds . . . to make negative points effectively."

In recent election cycles, presidential campaign advertising has increasingly emphasized the last of our four approaches: ads attempt to tie the choice of candidates to the preexisting value structure held by voters. For example, when Ronald Reagan was overwhelmingly reelected in 1984, many commentators stressed the importance of his vivid campaign commercials entitled "Morning in America." For much of the fall campaign, television viewers were cheered by a well-choreographed version of national resurgence, marked by vibrant color shots of the nation's heartland interspersed with beautiful pictures of a vigorous president. Despite the beauty and artistic grace of these communications, they hardly changed voter preferences.

According to Dr. Richard Wirthlin, Reagan's pollster, the campaign's victory margin was built upon two other commercials that are a good deal less memorable.[47] The first depicted a grizzly bear moving through a forest and arriving, at the end, in a face-to-face encounter with a native American; the second pictured three working Americans responding to the announcer's query as to how they would cope with Mondale's proposed tax increase. Though far less scenic, these commercials had the effect of establishing exactly the right cognitive links among values as prescribed by the campaign's research. If the "Morning in America" commercial is remembered as an atomic blast that blew away the Mondale campaign, these spots were precision-guided munitions directed by skilled marksmen.

Well before the fall campaign began, Wirthlin established a "cognitive map" of American voters based upon a small sample of in-depth personal interviews. The map allowed him to chart the conceptual relationships among firmly held values, preferences in terms of public policies or programs, and the campaign issues of the day. For example, at the deepest level of values, many Americans expressed the desire to remake the United States and the world into a better place for future generations. This overall goal had two dimensions: the value of preserving world peace and that of securing a better America.

Once this map was created, Wirthlin could assess the relative advantages of the two candidates. Large-sample surveys generated the data necessary to measure the beliefs held by probable voters as to which of the presidential contenders was more likely to secure these values. Unfortunately for Reagan's reelection prospects, at the outset of the campaign, American voters viewed Walter Mondale as more likely to avoid getting the United States into a war. Avoiding war was also very strongly associated with preserving world peace. In order to turn this value structure to Reagan's advantage, the campaign needed to raise the visibility of the deterrence argument: defense preparedness would lead to world peace.

At the time, many commentators considered the bear commercial ill defined and perplexing. Behind the scenes, however, careful monitoring established that the ad did achieve its purpose: focus groups demonstrated that viewers received the intended message that deterrence would preserve world peace. Reagan's campaign ran the bear commercial in critical states where undecided voters would be amenable to this value-based argument. Field surveys conducted in these markets measured a noticeable strengthening of the conceptual link between Reagan's policies advancing a strong national defense and the value of preserving peace in the world. By the end of the fall campaign, the majority of Americans viewed Reagan, not Mondale, as more likely to preserve world peace, even though they still saw Mondale as less likely to involve the nation in a war.

The second commercial was aimed at the domestic side of Wirthlin's value map. Three working-class Americans responded with skepticism and abuse when the announcer asked if they could afford Mondale's proposed tax increase. The ad endeavored to create cognitive links that worked to Reagan's advantage in a network of commonly expressed desires for a better America, for a secure future for oneself and one's children, and for economic recovery. Wirthlin argues that Americans receive information from commercials "through symbols, by tone, and by what people see, not through the words [used]." In thirty seconds this commercial managed to imply that the proposed tax burden would fall on those who could not afford it, thereby jeopardizing their future security. In effect, unfair taxes would undercut the economic security gained under the Reagan recovery. Focus groups documented that viewers gleaned this message from the ad, and by the end of the campaign, voter perceptions of competing candidates had changed in value terms. Even though most Americans still preferred many of Mondale's specific domestic policies, they viewed Reagan as the candidate more closely linked to their overarching value of securing a better America.

This does not mean that issues are absent from advertising directed toward values. In fact, the strategic process involves isolating issues that are both salient to voters and have consequences in terms of their preexisting values. Advertising achieves its impact by establishing a connection between these issues and their consequences in terms of values. Wirthlin summarized the way in which these ads worked: "You have to link an issue to a consequence [program] and drive to a value. That's when you get a punch in terms of the vote." Wirthlin went on to argue that the "Morning in America" commercial that received far more plaudits was, in fact, loaded with emotional content absent of any issue component. They were also ineffective as campaign communications.[48]

From the perspective of critics, however, campaign ads that appeal by emphasizing emotions, inherently depreciate rational processes that relate to the policy consequences of voting for one candidate or the other. As Kathleen Jamieson has observed,

> The notion that the end of rhetoric is judgment presupposes that rhetoric consists of argument—statement and proof. Morselized ads and news bites consist instead of statement alone, a move that invites us to judge the merit of the claim on the ethos of the speaker or the emotional appeals (pathos) enwrapping the claim. In the process, appeal to reason (logos)—one of Aristotle's prime artistic means of persuasion—is lost.[49]

Of course the issues used in campaign advertising can be either those of the character of the candidates or the policy issues facing the nation. In either case, appeals to emotions and values are a long way from the starting point of this discussion, namely the perceived effectiveness of providing information or making a persuasive argument. Instead of statement and proof, televised ads work primarily by heightening the visibility of some issues over others or by triggering the preexisting values of viewers.

The Problems of Political Advertising

Given the current dissatisfactions with our political system,[50] it should be no surprise that the central place of advertising in American politics is controversial. We will deal with three areas of criticism of contemporary campaign advertising: (1) the effects of negative or attack ads, (2) the quality of political discourse conducted through campaign commercials, and (3) the costs involved and related consequences. As might be expected, consultants dispute many of the implicit assumptions on which these complaints are based.

Probably the most significant complaint revolves around the

growing use of ads designed to weaken support for an opponent rather than to encourage voter support for the sponsoring candidate. The controversy over negative ads has emerged over the last decade and a half. While criticizing one's opponent is hardly a new invention, we might be concerned that the medium of television magnifies the potency and uncivil effects of negative campaigning. Even prominent political consultants such as Doug Bailey, David Sawyer, and Victor Kamber have begun to complain about the increasingly frequent recourse to negative ads.[51]

The term "negative ads" is used by critics of paid political advertisements precisely because it is so patently imprecise: "negative" in the sense of criticizing an opposing candidate or cause and "negative" in the sense of having harmful effects or being essentially unethical. Certainly ads that suggest reasons for not voting for a candidate, criticizing his or her character, policies, or record, need to be distinguished from ads that advocate a reason *for* voting for a particular candidate.[52] But we should demand precision from the critics of campaign advertising. Since the essence of an election contest lies in defining a choice for citizens to make, criticism of one's opponent is inherently necessary. Such criticism can be accurate or distorted, merited or unwarranted, ethical or unethical. That is, one problem lies in the *truthfulness* of the statements made in advocacy. Many of the complaints about negative ads fasten upon examples that have given voters a misleading or distorted impression of the opposition. The same objections should be valid about advertising that misrepresents the record of a candidate in order to gain him (or her) votes, but there is a sense that negative ads are inevitably misleading:

> When you attack the other guy in a thirty second ad, distortion almost has to be the rule. Because what political advertising does is find the one thing in a politician's career . . . that is outrageous to a majority of the people. And you play up that one thing. You blow it up all out of context.[53]

Two other charges are also leveled against negative ads.[54] First, they are said to increase voter cynicism and decrease turnout by discouraging voters about the choice they face. Critics especially worry that when both sides engage in sustained attacks, citizens will have a depreciated view of their options. Second, negative ads are criticized as resorting to innuendo and inference to make subtle appeals on the basis of race, color, creed, or gender. The infamous Willie Horton ad from the 1988 campaign is cited as a case in point.[55]

The first complaint is debatable, or at the very least researchable. While the charge is frequently made that negative ads increase voter cynicism, very little hard data exist to support the claim. There certainly are instances of advertising being created with the express purpose of raising doubts among an opponent's marginal supporters. Consider, for example, Doug Bailey's comments about a campaign in Ohio:

> We intentionally ran from day one until the end . . . extraordinarily negative ads against John Gilligan in Cleveland, but only in Cleveland, in order to drive turnout down, because . . . Cleveland would give Gilligan a plurality. And, in fact, if you look at the statistics, the turnout in Cuyahoga County was four or five points lower.

Obviously the candor of these remarks is rare, so that assessing the conditions under which the doubts created by advertising will translate into a higher rate of nonvoting cries out for more careful research rather than impressionistic charges. One experimental study by Garramone and her colleagues failed to detect any differences between positive and negative commercials in affecting voter turnout.[56] On the other hand, on the basis of preliminary analysis of research in progress, Shanto Iyengar reported a very slight decrease in voting among respondents exposed to negative ads.

The charge that voter turnoff can be considerable when both sides engage in heavy negative ads may be particularly difficult to sustain. These races tend to be accompanied by close competition and high spending, both factors that tend to increase turnout. Many of the contests in recent years that were notorious for negative ads from both sides—such as the last two races in which Jesse Helms sought reelection as U.S. senator from North Carolina—have also been characterized by fairly high voter turnouts.

This relationship is also complicated by the fact that politicians can turn the mounting cynicism of the public against their opponents. As political consultant Greg Schneiders observed,

> It is easier to drive negative numbers on the opponent than it is to build positive numbers for yourself, for a very simple reason. . . . [Y]ou start with a given: that most voters think all politicians are sleaze bags. [Then] to say, "my opponent is a sleaze bag," is going to hit a responsive chord. . . . It's the easiest thing in the world. If you're going to drive numbers in a hurry, that's the way to do it.

Turning to another complaint, negative ads may use innuendo and subtle cues that serve to enlist support by engaging presumptions and biases based on race, religion, or other characteristics that divide our society. This observation is hardly unique to televised advertisements criticizing the opponent. Most other forms of reaching voters may use these appeals including, for example, the most basic form of communication, word of mouth. On the other hand, do we find positive appeals for support that use these same cues to solidify a distinct community behind a candidate somehow less despicable? The complaint against television is really that visual ads directed toward voters in their living rooms are more powerful.

The complaint also serves to bring forth a latent contradiction in the workings of elections. From the viewpoint of campaigners, the fundamental task of building an electoral majority involves both generating cohesion and marking divisions within the pool of voters. Yet elections should provide the mechanism for collective choice that can resolve disputes peacefully. Consequently, appeals that become too divisive may energize so much group conflict that the essential purpose of elections is undercut. This concern leads us to ask political managers to exercise judgment in balancing the need to create divisions against the longer-run social consequences. But we are making an expansive request of those whose job is to win elections. To quote Lee Atwater, who ran George Bush's 1988 campaign: "We had only one goal, to help elect George Bush. That's the purpose of any political campaign. What other function should a campaign have?"[57]

The heavy reliance on thirty-second commercials has generated another general lament over the resulting quality of political rhetoric. Paid ads, it is said, do not really serve the public good as an instrument of robust debate and discussion. This complaint often becomes diffuse because, as with many important aspects of campaign communication, we cannot really separate the efforts of politicians to reach voters through news reporting from their advertisements, which are more clearly under their control.[58] The same criticism can just as legitimately be directed toward television news programming, which has steadily reduced the average length of time in which candidates appear on the screen.[59]

From the viewpoint of political communicators, the fact that advertising time is sold primarily in thirty-second packages is part of the problem. Obviously very little can be said in the allotted time, leading to an impoverishment of ideas advanced through the campaign's primary vehicle. We should not, however, accept too readily this defense offered by political consultants. To be sure, if the media producers were to develop longer commercials, they would face a protracted struggle getting broadcast licensees to clear the time for them. But ultimately the campaigners would probably prevail.[60] Rather than adding this fight to an already conflicting relationship, political managers would rather go along with the routines of commercial broadcasters.[61] We should also realize that the five-minute "bio" ads produced by many campaigns are not greatly superior in the depth of information conveyed about the candidate and her (or his) policy proposals.

Finally, both campaigners and broadcasters express reservations about the audience's role in the processes of political communication. Both communities voice considerable alarm over the growing tendency toward "zapping." With more channels to choose from and remote controls, impatient viewers are reportedly switching channels at the slightest excuse. Since political programming is not overwhelmingly popular, both communities worry that longer ads will simply be zapped into oblivion.[62]

This observation raises a broader point about the role of elections and our expectations for the communication process, which is related to our discussion of whether ads should work to educate voters. Those who are highly critical of campaign ads often argue that they are sparse on details and present voters with very little useful information. In addition to arguing that they have neither the time nor the resources to educate voters on the details of public policy, campaigners sometimes contend that voters in general show little demand for or interest in these complexities.

Social science can shed some light on this dispute. The debate begun in the 1960s over the extent of rationality in voter thinking has recently shifted in favor of a greater respect for their abilities and willingness to process relevant information. Instead of debating the level of ideological sophistication or the degree of coherence in voter belief systems, political scientists are now arguing over the extent to which voters make rational calculations, on the one hand, versus the conditions under which they might even become deliberative, on the other. Most campaigners would accept Samuel Popkin's argument that voters can and do process rudimentary information and cast their ballots based on calculations of self-interest.[63] While many ads do have a strong emotional appeal, their effectiveness depends upon some level of rationality in voter decision making. Fishkin and those who hope for a more elevated standard of political discourse believe that voters are capable of more.[64] They view the problems of contemporary elections as institutional in nature and wish to underwrite mechanisms through which voters can receive more information, debate alternatives with their peers, and come to a more deliberative decision. Presumably campaign communications and advocacy will play an important part in these institutions, but not in the form of paid thirty-second commercials.

Concern about the quality of the public discourse during campaigns has another facet: in the aftermath of the 1988 election, numerous complaints surfaced that advertising tended to focus voter attention on the less important issues facing the country. In speeches and ads, George Bush attacked his opponent for his positions on the pledge of allegiance, the American Civil Liberties Union, and furloughs for serious offenders. Less attention was paid to more important issues such as the mounting federal deficit and our relations with the deteriorating Soviet Union. As media consultant David Sawyer opined, "Our political campaigns are out of control. They don't even *raise* the vital issues that our government must address." [65]

Note that this objection, which is only partly directed at campaign ads per se, is really a general indictment of contemporary campaigning. Yet, even though these campaigns may be fought out on spurious grounds, the charge should be embraced rather cautiously on two grounds. First, the whole purpose of election contests is to establish a contained struggle over what is important for voters to consider in making their choices. Of course, that which is central from one perspective may be trivial from another. Moreover, there is a measure of accuracy in the defense offered by Bush's campaigners that their topics of debate were merely symbols, intended to raise questions about Dukakis's judgment and basic political instincts. The second reason for setting aside the complaint that ads trivialize political debate is that the possible cures are all worse than the objection itself. If free speech enshrined in the First Amendment means anything, it means protection against efforts to place specific political topics out of bounds.

A third and final complaint about the extensive reliance on television by modern presidential campaigns concerns the costs involved. Since paid television constitutes the largest single item in the campaign budget, we are inevitably led to a discussion of campaign financing. As this subject cannot be fully covered here, however, we can do little more than flag these issues for discussion elsewhere.

The need to pay for advertising campaigns has become the subject of extensive debate over the public policies shaping presidential campaigning. It is certainly true that the huge percentage of their campaign treasuries that candidates pour into television advertising is largely responsible for the sharp increase in the overall costs of presidential campaigning over the past three decades. As noted above, Alexander and Bauer report that between 50 and 70 percent of the general election dollar went into television.[66] Overall, they estimate that all candidates for president in 1988 spent a total of $500 million.

Arguments over money in politics lead in several directions. Some observers are troubled that too much money is spent and much of it is wasted. Others express the concern that the percentage allocated to television advertising is diverted from funding for grass-roots activities that encourage direct citizen participation in the campaign. Some critiques fasten upon the quid pro quos that politicians must give to extract money from potential contributors; in effect, they see contributions as synonymous with bribes. The same argument can take a more subtle form: even if candidates do not make specific pledges to contributors, the requirements of raising money necessarily divert the candidates' attention from the mass of voters toward communities that can provide substantial contributions.[67]

The most fundamental of these questions is how much money is justified in politics. On one side, it is quite clear that, operationally, a great deal of the money spent by presidential candidates is wasted. But since this becomes clear only in retrospect, it cannot be used as a justification for reducing the level of resources available to candidates. Political managers also have a good point when they argue that their spending on television is modest in comparison with that of commercial advertisers. For example, during a nonpresidential election cycle, all candidates for House and Senate combined spent less than 20 percent of the TV advertising budget of one company (Proctor & Gamble) in that same year.[68] More polemically, consultants ask whether 87¢ per general election vote is too much to spend on the selection of our nation's leader. When addressing the presidential level specifically, they also note that, in imposing very sharp limitations, the current law reduces unduly the amount that a candidate can allocate to any one state, especially if the appropriate comparison is to spending for gubernatorial or senatorial contests in that same state. On balance, they appear to have the better of the argument, at least in specifics.

There is also an effective answer to the complaint that campaigns invest heavily in television advertising to the detriment of spending that would enhance grass-roots participation by voters. We should recall that the Federal Election Campaign Act was amended in 1979 in response to these complaints.[69] The result has been an explosion of funds directed into state party treasuries specifically reserved for grass-roots and party-building activities. These are the so-called soft-money accounts, which have risen dramatically in terms of the amounts collected and dispersed as well as in complaints about this loophole in the law.[70]

This reminder from recent history provides a perfect segue into our last topic: what can be done about the problems we perceive in campaign advertising. In turning to a consideration of policy responses to paid political advertising, the example of soft money should help us recall that legislation often has unforeseen consequences.

What Can Be Done

If campaigns are increasingly negative and divisive, if they are not giving voters the information necessary to reach appropriate judgments on national leadership, if they cost a great deal, if the need to raise money has detrimental consequences, in short, if modern presidential campaigns are not working to strengthen American democracy, what can be done about it?

The proposed remedies can be examined from a number of perspectives. Even though a full exposition of the potential solutions is beyond the scope of this summary essay, I will examine proposals for alleviating the detrimental aspects of television advertising from the vantage point of four theaters in which these matters might be resolved: congressional legislation, the contributions of journalists, the creation of citizen commissions to critique the conduct of campaigns, and the prospects for self-regulation by those who create presidential television.[71] Of course the forum within which remedies are sought and alternatives offered depends largely upon one's diagnosis of the malady.

Legislation

A problem of this scope and visibility has naturally drawn the attention of those who exercise statutory power. Especially given the important role of public funding in presidential campaigns, Washington policy makers have attempted to address these concerns through legislation.[72] Since these same problems occur on the congressional level, many of the resulting proposals are entangled with potential reforms in the laws regulating campaign finance of all federal candidates.

Not surprisingly, little in the way of consensus can be found in the numerous proposals advanced for circumscribing political advertising. Some critics even demand that paid political advertisements be banned entirely from television; others would regulate the format of political broadcasting by mandating that stations sell time to political advertisers only in longer segments (three to five minutes); still others would require changes in the content of televised commercials. For the purposes of this essay, a wealth of legislative proposals can be summarized as taking one of three approaches.[73]

As campaigns have turned more negative, the public has become increasingly distressed about their conduct. According to a recent survey, only 38 percent of the respondents considered campaign commercials "somewhat or very useful," while 59 percent said they were "not useful at all." [74] Heavy criticism has led many to question the need for political advertising. "Why should candidates be sold like soap?" they ask.

Those who propose that television ads be eliminated do understand the critical importance that electronic communications play in a modern polity.[75] Accordingly, they concentrate on alternative means that would allow politicians to converse with voters on a more edifying plateau. Over the years, numerous suggestions have been advanced that would require broadcast licensees to provide free time, usually in half-hour segments, as a condition of their licensed use of the public airwaves. Politically, however, crafting a specific proposal that can embrace the differences in the pattern of overlap of media markets and constituencies has proved elusive. For stations located in major metro-politan areas, providing free time to all candidates within their market could preempt a large percentage of prime-time hours in the weeks before the election. Conversely, if candidates were given a voucher for a set amount of broadcast time, some congressional contestants would be able to reach all of their potential constituents, while others would reach only a fraction.

More recently, proposals have focused on requiring candidates to address voters in specified formats in return for public funding of their campaigns. For example, at least three legislative proposals would require these candidates to participate in debates sponsored by a nonpartisan or bipartisan organization. Another would mandate that presidential candidates take part in six weekly informational broad-casts, each of which would address a single policy question.[76]

Another approach concentrates its indictment of political ads on the level of information that can be communicated in fifteen or thirty seconds. One piece of suggested legislation would make the lowest unit rate provision of the Communications Act apply only to political ads of one minute or longer. Another suggestion would provide vouchers to reimburse Senate candidates for their expenditures for ads of one to five minutes in length. Presumably the current formats are too brief to allow a substantive presentation of complex policy matters. The authors of such legislation assume that the present form of highly abbreviated political messages lends itself to attacks on one's opponent and to emotional and distorted appeals. That sixty-second or five-minute commercials would provide more substantive information is, however, an unproven and perhaps unwarranted assumption.

A more aggressive approach attempts to regulate the content of ads in one form or another. Some reformers would stiffen the requirements through which sponsors of ads are identified, requiring an image of the candidate and an audio statement by him or her authorizing the ad. In addition, the size of the statement and the length of time it must appear on the screen would be specified in some detail. Other proposals would require ads to adopt a "talking heads" format; the candidates them-selves would have to appear throughout the ad. Presumably, as Jack Winsbro notes, "few candidates are at all likely to be completely negative when they are delivering the message themselves."[77] One major reform package, which reached the Senate floor in 1987 but died in a filibuster, would have gone further, prohibiting the use of "staged reproductions of any event or scene in Senate campaign commercials." [78] Still another idea would require candidates to make any reference to an opponent in person; otherwise that opponent would receive free compensatory air time in which to respond.

Despite the multitude of proposals, these efforts to regulate political speech through governmental action will not be accomplished easily. All of them face stiff legal challenges.[79] At their core, First Amendment protections apply to political speech during an election. As the Supreme Court argued in *The New York Times v. Sullivan,* "debate on public issues should be uninhibited, robust, and wide open, and that may well include vehement, caustic, and sometimes unpleasantly sharp attacks on government and public officials." [80] Nevertheless, in *Buckley v. Valeo,* the Court provides an exception to its blanket rejection of expenditure limitations as an unconstitutional restraint on speech. Limits are acceptable in presidential campaigns in which voluntary public funding is involved. This logic has given some hope to the proponents of measures to restrain what they see as unfortunate abuses in paid political advertising.

The Role of Journalists

Over the past decade, numerous other critics have advanced reform proposals that stop short of legislation. Among these are a more aggressive approach in campaign journalism, the establishment of a fair campaign practices commission, and a stronger code of ethical conduct agreed to by professional advertising consultants. All of these would rely to some extent on the force of public criticism to contain the manipulative impulses of campaigners.

In recent elections many news outlets have instituted "ad watch" coverage, reporting on the ads that candidates are using and the claims they are making in their ads. In many cases these stories have unquestionably contributed to the impact of the ad itself, for example, the infamous Willie Horton ad created during the 1988 campaign by an independent expenditure committee. The ad received a limited run on cable systems in Maryland, but that was never its real purpose. Larry McCarthy, the producer, passed a copy to the producers of the syndicated talk show "The McLaughlin Group." Once the ad was aired on that program, it was picked up by the television networks and run on their news broadcasts.[81] News reporting greatly magnified the reach and presumably the impact of the ad.

The ad-watch programs conducted by numerous newspapers and a few television stations constitute a more assertive form of news reporting of political commercials. Journalists attempt to analyze the ads and to disclose the veracity of their assertions and the subtle messages the ads seek to convey. The import of this reporting is presumably to inoculate voters against emotional manipulation and to make candidates more responsible when leveling charges against their opponents. Though this reporting is admittedly in its infancy, success has thus far been mixed. The resources necessary to check the accuracy of assertions made in the ads are often lacking, forcing journalists to rely on documentation provided by the candidates. Rarely do journalists have access to databases matching those of the national committees of both parties, which compile detailed records of the words, votes, and actions of possible opposing candidates. Moreover, an effective ad critique requires more in the way of judgment than a detailed set of records. Consider a typical statement such as "he voted to increase your taxes twenty times in the last four years." The consultant producing this commercial may indeed point to twenty votes, but journalists have to go back to the legislative history to judge how clearly some of those votes could accurately be categorized as votes to increase taxes.

Furthermore, ascertaining the accuracy of statements made in television ads constitutes only one aspect of potential criticism. Understandably the spoken and written words contained in advertising are vastly easier to analyze, but, as we have learned, much of the impact of advertising is conveyed by pictures, lighting, background music, and so on. Pulling these aspects apart is a more subjective undertaking. In addition, without the strategic information necessary to understand the impact of the ad, it is very difficult for journalists to comprehend fully its purpose. Imagine, for example, an ad-watch piece on Reagan's "Bear in the Woods" commercial. The possibility that reporters could diagnose the information contained in that spot is remote.

Though a vigorous press might engender some caution among presidential candidates and their consultants, we suspect that the restraining role of journalists will never become robust enough to fully check the potential for abuse. For one thing, the journalists' vigor in exposing such offenses will certainly be circumscribed by their reliance on politicians as sources of information. In order to do their work, journalists need access to presidential candidates and their immediate advisers. The all too frequent use of unattributed quotes and off-the-record briefings serve to undercut an energetic adversarial approach to candidates and consultants. Relations between reporters and politicians who provide them with stories can at times become symbiotic.[82] In recent years, moreover, many news organizations have functioned under very severe financial constraints that may militate against devoting the research time necessary to screen the accuracy of televised commercials. The traditional pressures of working against deadlines and limitations of news coverage are further deterrents to allocating the time and resources necessary to grapple with abuses other than the most egregious.

The canon of objectivity can also serve to make journalists wary about taking on the self-assigned truth-squad function.[83] To be sure, the legacy of Watergate may have accentuated the prominence of investigative journalism, but in a campaign context journalists are highly aware that criticizing one candidate will assist the opponents. If only one candidate is running scurrilous ads, taking on that candidate alone will upset the natural balance in campaign coverage that journalists prefer.

For these and other reasons, we should not rely too heavily upon journalists to provide an effective check on misleading and deceptive advertising. Nevertheless, despite its shortcomings, ad-watch reporting has had a worthwhile impact to the extent that it has made campaigners slightly more cautious about making unsubstantiated charges. If such reporting only serves to alert campaigners to the need to compile documentation for the assertions contained in their advertising, it should be encouraged. Ad-watch programs, however, are not a cure-all.

Private Associations

Another institutional method for containing abuse in political advertising might be found in voluntary organizations using the power of public opinion to condemn egregious breaches of ethical practice. On the local level, numerous committees have been organized to contain campaign abuses.[84] The most ambitious and long term of these efforts on a national level, the Fair Campaign Practices Commission (FCPC), functioned between 1954 and 1975. A prominent roster of civic and political leaders asked candidates to sign a Code of Fair Campaign Practices, agreeing to abstain from false, distorted, or unsubstantiated campaign materials, racial or religious appeals, and character defamation. The commission published the names of those who refused to take the pledge and also investigated cases of alleged abuse brought before it—an average of sixty-three complaints in each congressional election cycle.[85] A finding by the commission that a campaign had exceeded the bounds of fair play would, it was hoped, elicit news coverage and thus public attention, allowing voters to reach decisions on the gravity of the matter. Even without an effective legal sanction, the FCPC purportedly did have some impact.

During the 1960s and early 1970s, the FCPC gradually curtailed its activities because it lacked a secure financial base. The funding that it was able to achieve received a severe blow in 1976 when the Internal

Revenue Service (IRS) ruled that publication of the names of candidates who refused to sign the code constituted an attempt to influence the outcome of an election and thus violated the FCPC's tax-exempt status as a nonprofit organization. Attacked by some as a self-appointed group seeking to replace the electorate and cut off from a stream of private contributions, the commission effectively ceased functioning in 1976 shortly after the creation of the Federal Election Commission (FEC). Though the problems of political rhetoric and the regulation of campaign finance are very different questions, in some ways this history has served to link these two ethical questions. In the sad aftermath of Watergate, problems of appropriate conduct in politics have been perceived primarily as a matter of campaign financing. Finance reform has eclipsed the work of the FCPC, even though money in politics and problems of rhetoric and advertising are distinct issues.

Unless the IRS is prepared to take a different view of these activities, re-creating a commission based on voluntary contributions is unlikely. Efforts to establish such monitoring commissions with public funds, moreover, are unlikely to withstand legal challenges. First Amendment values too thoroughly pervade political speech of all kinds. For example, after the demise of the FCPC, the New York State Legislature directed the Board of Elections to enforce a code of fair campaign practices; a federal court invalidated the action, warning of its chilling effect on political speech.[86]

Conditions have, nevertheless, changed markedly in the thirty-seven years since the creation of the FCPC. Citizens have become more cynical about the functioning of democratic politics, and the outright abuses and questionable practices have, if anything, become more visible and possibly more widespread. Perhaps most dramatically, the small cadre of political consultants has expanded into an industry that is increasingly a focal point of complaints. Calls for reestablishing a campaign practices commission have recently been rising.[87]

Voluntary Restraint

It is always possible that voluntary restraint by those involved could become a means of containing abuse in campaign advertising. Voluntary agreements among opposing candidates to restrain their campaign in rhetoric or spending are not unknown. Particularly during the nomination stage, the national chairmen of both political parties have attempted to cajole candidates to refrain from damaging one another. Using behind-the-scenes pressure and public statements, they have had mixed success in limiting vituperative attacks among their party aspirants.[88] In 1989, the two national chairmen reportedly even agreed to a joint conference to discuss ways of curbing campaign abuse during the 1992 general election campaign.[89] The meeting never took place, most likely because of Atwater's subsequent illness.[90] Nevertheless, the fact that these two felt compelled to address the problem collectively, even if they only considered it a problem of public relations, means that all the criticism of the 1988 campaign had an impact on the practitioners.

Almost inevitably such agreements last only until the prospects of winning and losing become clearer. The emergence of a front-runner usually signals the escalation of rhetoric in the attempt to even the competition.[91] Without an institutional structure that can hold candidates and consultants accountable to the agreements made before the race begins, when everything still seems possible, these individual voluntary agreements will most likely flower in the spring, only to wither in the fall.

Farther down the road, an emerging notion of professionalism among political managers might even make it possible to create the institutional grounding necessary to reinforce self-restraint. The American Association of Political Consultants (AAPC) has, since its inception, advocated a code of ethics to its membership, administered by a standing committee. But, as the association's officers are the first to admit, it is virtually impossible to enforce a code on members who can easily withdraw from membership without substantial damage to their business. The whole topic of ethics in political consulting has not been given much discussion by practitioners, so that at present their community lacks anything approaching a consensus on appropriate or responsible conduct. Nevertheless, the AAPC has recently begun to reexamine its code. In March 1991 a membership conference was held to discuss four major questions, including those surrounding campaign television commercials.[92] The stream of postconference memos and proposals suggests that the conference may have initiated what should predictably be a lengthy discussion.

In the final analysis, it appears unlikely that any of these institutional vehicles—legislation, journalism, private associations or voluntary restraint—will have a lasting impact as long as public debate over these problems remains impoverished. As Robert Fullinwider has observed:

> Such an arid environment gives no nourishment to the emergence of a culture of ethical reflection and discussion among political professionals about the means and ends of politics. In the absence of such a culture, external checks on campaigns serve mainly as obstacles to be overcome by clever political managers, who remain complacent about their activities.[93]

Ads and their creators have clearly become a subject of greater controversy. For example, whether or not he produced the Willie Horton ad, Roger Ailes clearly did become exposed to criticism and attack after his ad campaign for Bush. In the following year, when he signed on to one side in a referendum battle over a proposed new Denver airport, the local newspapers made an issue of his participation. In other words, to a small degree, self-corrective processes may exist in this business.[94]

Criticism in one isolated case, however, is unlikely to achieve any lasting impact. Moreover, it is hardly fair to single out an individual practitioner. The climate of discussion of these problems by the public at large remains very far from the level of intensity and complexity required to influence the advertising operations of presidential campaigns. In short, we need much more discussion and airing of the weaknesses and strengths of the present system. Once we understand the problem, however, there is little to be gained in a discussion that condemns the whole undertaking of modern political communication as irredeemable while making no effort to comprehend the very real needs and limitations of political communicators.

Notes

1. Daniel J. O'Keefe, *Persuasion: Theory and Research* (Newbury Park, Calif.: Sage, 1990), 200-223.
2. See Edward Tufte, *Political Control of the Economy* (Princeton, N.J.: Princeton University Press, 1978), or Stephen Rosenstone, *Forecasting Presidential Elections* (New Haven, Conn.: Yale University Press, 1983).
3. Steven J. Rosenstone, quoted by Richard Morin, "The Real Secret of 1988: Why Negative Political Ads Didn't Much Matter," *Washington Post*, February 12, 1989, C1-2.
4. Ethiel Klein, personal communication, March 1992.
5. Any attempt to define the total pool of presidential candidates immediately encounters the problem of delineating serious candidacies. My count, which is certainly not all inclusive, tallies seventy-five major aspirants from 1960 to 1988. The Democratic candidates include the following: in 1960: Estes Kefauver, John Kennedy, Lyndon Johnson, and Adlai Stevenson; in 1964: Lyndon Johnson; in 1968: Hubert Humphrey, Robert Kennedy, Lyndon Johnson, and Eugene McCarthy; in 1972: Shirley Chisholm, Hubert Humphrey, Henry Jackson, George McGovern, Wilbur Mills, Edmund Muskie, Terry Sanford, and George Wallace; in 1976: Birch Bayh, Jerry Brown, Jimmy Carter, Frank Church, Fred Harris, Henry Jackson, Eugene McCarthy, George McGovern, Milton Shapp, Sargent Shriver, and Morris Udall; in 1980: Jerry Brown,

Jimmy Carter, and Edward Kennedy; in 1984: Reubin Askew, Alan Cranston, John Glenn, Gary Hart, Ernest Hollings, Jesse Jackson, George McGovern, and Walter Mondale; in 1988: Bruce Babbitt, Michael Dukakis, Albert Gore, Richard Gephardt, Gary Hart, and Paul Simon. The Republican candidates include the following: in 1960: Richard Nixon; in 1964: Barry Goldwater, Nelson Rockefeller, and George Romney; in 1968: Richard Nixon and Nelson Rockefeller; in 1972: John Ashbrook, Paul McCloskey, and Richard Nixon; in 1976: Gerald Ford and Ronald Reagan; in 1980: John Anderson, Howard Baker, George Bush, John Connally, Philip Crane, Robert Dole, Benjamin Fernandez, Larry Pressler, Ronald Reagan, and Lowell Weicker; in 1984: Ronald Reagan; in 1988: George Bush, Robert Dole, Pierre du Pont, Alexander Haig, Jack Kemp, and Pat Robertson. Of course, not all of these candidates were able to mobilize the resources necessary to produce the broadcast television ads.

6. Herbert E. Alexander and Monica Bauer, *Financing the 1988 Election* (Boulder, Colo.: Westview Press, 1991), 36 and 19.

7. According to data released by the Federal Election Commission, in January 1991 the voting age population of the United States was 185,105,441, not including American Samoa, Guam, Puerto Rico, and the Virgin Islands.

8. Of course, whether nonvoters differ dramatically from voters in their political attitudes, predispositions, or candidate preferences is a highly debatable question. To the extent they do differ, the messages crafted by campaigners on the basis of their polls will do very little to energize turnout.

9. As I shall note below, however, 5 percent can be a significant proportion of the undecided voters. In essence, campaigners do not really dispute that organization can be a useful tool, all other things being equal. Five percent, after all, may decide the election. Rather, their calculations become those of the efficient use of scarce resources. Even if one can rely upon a stream of volunteers, the costs and management energy required to recruit, train, equip, and direct campaign workers to make contacts of the magnitude needed in presidential campaigns can be substantial. Campaign managers are confronted by a host of potential consultants, each arguing that his or her particular specialty will be decisive. Television advertising has the appeal of being manageable by a small group of people and being perceived as more cost efficient in terms of contacts per dollar spent.

10. Edward J. Rollins, campaign director for Reagan-Bush in 1984, speaking to class at The Graduate School of Political Management on November 11, 1991, covered by C-SPAN.

11. Ibid., and Robert Beckel, campaign manager for Walter Mondale in 1984, speaking to class at The Graduate School of Political Management on November 11, 1991, covered by C-SPAN.

12. The percentage of popular votes garnered by the major candidates over

this period is as follows. (The names of the winners appear in all capital
letters.)

Election year	Democratic		Republican		Independent	
	Candidate	%	Candidate	%	Candidate	%
1960	KENNEDY	49.7	Nixon	49.5		
1964	JOHNSON	61.0	Goldwater	38.5		
1968	Humphrey	42.7	NIXON	43.4	Wallace	13.5
1972	McGovern	37.5	NIXON	60.7		
1976	CARTER	50.1	Ford	48.0		
1980	Carter	41.0	REAGAN	50.7	Anderson	6.6
1984	Mondale	40.6	REAGAN	58.8		
1988	Dukakis	45.6	BUSH	53.4		

SOURCE: Congressional Quarterly, *Guide to U.S. Elections,* 2nd ed.
(Washington, D.C.: Congressional Quarterly, 1985), and Richard
Scammon and Alice McGillivray, *America Votes,* vol. 19 (Washington,
D.C.: Congressional Quarterly, 1991).

13. For an introduction to the manifold voting studies that address the concept
of a normal party vote, see Eric R. A. N. Smith, *The Unchanging
American Voter* (Berkeley, Calif.: University of California Press, 1989);
and Samuel Popkin, *The Reasoning Voter: Communication and Persuasion in Presidential Campaigns* (Chicago: University of Chicago Press,
1991).

14. The average percentage deciding before the conventions is 45 percent for
the years 1956, 1964, 1972, 1976, 1980, and 1984, and only 33 percent for
other years. These data are even more striking in that the data in Table 3-
1 are recorded from the original responses, where a high percentage of
respondents actually chose the answer "Knew all along."

15. For discussions of retrospective voting, see Morris P. Fiorina, *Retrospective Voting in American National Elections* (New Haven, Conn.: Yale
University Press, 1979).

16. Among "Independent-Democrats" and "Independent-Independents,"
claims of making the vote decision during the campaign are quite often
upward of 50 percent and even into the sixtieth percentile. By contrast,
fewer "Independent-Republicans" report deciding during the campaign.
But these figures contrast markedly with strong party identifiers, among
whom the majority respond that they knew all along or they decided when
the candidate announced. See Warren E. Miller and Santa A. Traugott,
American National Election Studies Data Sourcebook, 1952-1986 (Cambridge, Mass.: Harvard University Press, 1989), 320. For a new analysis
arguing that growth in the number and importance of independent voters
has been vastly overstated, see Bruce E. Keith, David B. Magleby,
Candice J. Nelson, Elizabeth Orr, Mark C. Westlye, and Raymond
Wolfinger, *The Myth of the Independent Voter* (Berkeley, Calif.: University of California Press, 1992).

17. Shifts of this nature would particularly affect those surveys whose "screen" for likely voters is based upon a simple question, "Do you think you will probably vote in the upcoming election?" Most surveys use multiple questions including behavioral items such as whether the respondent reports actually voting in the last election.

18. Here we should take note of the very different tasks of political analysts and political managers: while the former may have the luxury (indeed the duty) of pointing out uncertainties in our knowledge of the voting process, those who take on the job of winning elections must implement strategy in spite of these uncertainties. Moreover, since their own effectiveness depends upon persuasion within the campaign team, political managers must appear confident and certain of the steps necessary to achieve victory, even where they are shrewd enough to recognize the frailty of knowledge upon which they act.

19. Jonathan Robbin, "Geodemographics: The New Magic," *Campaigns and Elections* (Spring 1980): 25-45; and Joseph Mockus, "Geodemographics II: Targeting Your Turnout," *Campaigns and Elections* (Summer 1980): 55-63.

20. Christopher Arterton, "Communication Technology and Political Campaigns in 1982: Assessing the Implications," a paper published by the Roosevelt Center for American Policy Studies, Washington, D.C. (June 1983).

21. Partick Barwise and Andrew Ehrenberg, *Television and Its Audience* (Newbury Park, Calif.: Sage, 1988), 12-48; and George Comstock, *Television in America,* 2nd ed. (Newbury Park, Calif.: Sage, 1991), 81-109.

22. Comstock, *Television in America,* Chapter 3.

23. This fact was pointed out to me by Democratic pollster William Hamilton, personal interview, March 1992.

24. Technically a gross rating point (GRP) is that percentage of the viewers within a particular market that is reached by a given television program as determined by either A. C. Nielsen or Arbitron. Because they are interested in reaching only voters, most media buyers for presidential campaigns prefer to work with adult gross rating points (AGRP). A "buy of 200" means purchasing enough thirty-second commercials that the rating points for those shows add up to 200, or 200 percent. But since individuals differ dramatically in their habits of watching television, one cannot assume that everyone in the media market will have seen the ad twice. Those who watch a lot of television will see it more often. One could, for example, buy lots of repeats between midnight and 7 a.m. Eventually the buy will tally 200 and the candidate might do well with the insomniac vote, but little else.

25. Private communication from Douglas Bailey, president of Bailey-Deardourff reporting on a race for the U.S. Senate in 1986 in North Dakota.

26. For a discussion of the consequences for electoral politics of these ongoing

changes in the structure of communications media, see Jeffrey Abramson, F. Christopher Arterton, and Gary R. Orren, *The Electronic Commonwealth: The Impact of New Media Technologies on Democratic Politics* (New York: Basic Books, 1988), especially Chapter 3.

27. Technically candidates for federal office do not have to provide broadcast stations with any documentation whatever for the claims they make in their ads. However, given the recent development of "ad watches" conducted by journalists probing the truth or falsity of political commercials, many campaigns have begun to collect and make available at least some substantiation.

28. For example, in 1988, the Eagle Forum headed by Phyllis Schlafly produced a short documentary entitled "Justice on Furlough," which attacked the Massachusetts prison furlough program and repeatedly used pictures of Willie Horton and his victims. Because the video did not mention that Michael Dukakis was running for president and did not therefore advocate his defeat or election, the production of this video fell outside the reporting requirements of the 1974 Federal Election Campaign Act (as amended).

29. Kathy Ardleigh, vice-president of The Media Team, during presentation at The Graduate School of Political Management, February 6, 1992.

30. Victor Kamber describes a textbook campaign of four cycles for building name recognition, establishing credibility, attacking the opponent, and returning to a positive tone in "feel-good commercials." See "Trivial Pursuit: Negative Advertising and the Decay of Political Discourse" (Washington, D.C.: The Kamber Group, mimeo, March, 1991).

31. Thomas Edmonds, president of Edmonds Associates, in presentation to The Graduate School of Political Management, June 1990.

32. I will ignore, for the moment, the critical debate over the propriety of contemporary politics that is inherent in this double entendre: many critics would observe that for the politicians the real question is "What do I *have* to say in order to get elected?"

33. See, for example, the work of Edwin Diamond and Steven Bates, *The Spot: The Rise of Political Advertising on Television*, rev. ed. (Cambridge, Mass.: MIT Press, 1988); Kathleen Hall Jamieson, *Packaging the Presidency: A History and Criticism of Presidential Campaign Advertising* (New York: Oxford University Press, 1984); and Montague Kern, *Thirty-Second Politics: Political Advertising in the 1980's* (New York: Praeger, 1989).

34. Karen S. Johnson-Cartee and Gary A. Copeland, *Negative Political Advertising* (Hillsdale, N.J.: Lawrence Erlbaum Associates, 1991). Although their study is confined to negative ads, in many places the discussion is relevant to all forms of political advertising.

35. Ibid., 140-161.

36. In a review of the communications literature on the production of persuasive messages, O'Keefe concludes that it is much too premature to aspire to dependable generalizations as to how media producers go about

their work. *Persuasion,* 216-220.

37. This is particularly true in the general election phase of presidential campaigns where, thanks to public funding, the amount of obtainable resources is known for each campaign, available at the beginning of the race and ample enough to make research and production costs a small percentage of the overall media budget.

38. In both 1980 and 1984 the Reagan campaign bought time in a few isolated markets so that surveys could be conducted before and after the ads ran. Personal interview with Richard Wirthlin, January 1987.

39. Strictly speaking, tracking polls involve continuous daily surveying of opinion in which an adequate sample results from combining data gathered on three or four successive evenings. As each new daily measurement is added, the data from the oldest evening is dropped off, creating a rolling sample. Tracking polls will often be started just before the campaign initiates its ad program. Then the effects of introducing a new spot or a sequence of ads can be observed by monitoring opinion change while the ads are on the air. Too frequently, however, campaign consultants will refer to separate cross-sectional samples conducted before and after a run of ads as their "tracking polls." But this usage is designed to give a mundane pre-and-post design the aura of an expensive and much desired technique.

40. Even when academics make the effort to translate communication theories onto a level that would be useful for practicing communicators, their perspective stands in the way. For example, Kathleen Reardon endeavors to summarize much of the persuasion literature, from classical theories to the latest debates, for the benefit of academics and practitioners alike. Her chapter on political persuasion discusses cognitive consistency theory, the rational choice approach, the cognitive schema perspective, and her own model, which assesses the perceived appropriateness, consistency, and effectiveness of the persuasive message. In short, she deals out the best cards that academia has to offer, but when they arrive in the hand of the political consultants, they are all blanks. See Kathleen Kelley Reardon, *Persuasion in Practice* (Newbury Park, Calif.: Sage, 1991).

41. Remarks of Mark Crispin Miller, quoted by Ellen Hume, "Restoring the Bond: Connecting Campaign Coverage to Voters," in *Campaign Lesson for '92,* a report of the Campaign Lessons for '92 Project (Cambridge, Mass.: Joan Shorenstein Barone Center, Harvard University, November 1991), 93.

42. As far back as the 1972 campaign, Patterson and McClure found that advertising in a presidential campaign conveyed a great deal of information. At the time, their research caused a stir because they concluded that ads communicated more information relevant to the election than did television news programming. See Thomas Patterson and Douglas McClure, *The Unseeing Eye* (New York: Putnam, 1976).

43. For a discussion of the potency of news media in agenda setting, see Shanto Iyengar and Donald R. Kinder, *News That Matters* (Chicago:

University of Chicago Press, 1987).

44. Diamond and Bates, *The Spot,* point out that commercials designed to erode support for the opponent have been used since the beginning of political spots. Even so, until 1978, when the National Conservative Political Action Committee (NCPAC) mounted an extensive campaign against several incumbent Democratic senators, the presumption within the industry was that these attacks had to be undertaken with great caution. Since 1978, the recourse to attack ads has become easier and much more frequent.

45. Research by Gary A. Copeland and Karen Johnson-Cartee discovered that voters react differently to these two types of ads; see "Southerners' Acceptance of Negative Political Advertising and Political Efficacy and Activity Levels," *Southeastern Political Review* 18 (Fall 1990): 85-102. These findings support those of B. L. Roddy and Gina M. Garramone, "Appeals and Strategies of Negative Political Advertising," *Journal of Broadcasting & Electronic Media* 32 (Fall 1988): 415-427.

46. Gina M. Garramone, Charles K. Atkin, Bruce E. Pinkleton, and Richard T. Cole, "Effects of Negative Political Advertising on the Political Process," *Journal of Broadcasting & Electronic Media* 34 (Summer 1990): 299-311; and Richard A. Joslyn, "Political Advertising and the Meaning of Elections," in Linda L. Kaid, Daniel Nimmo, and K. R. Sanders, eds., *New Perspectives on Political Advertising* (Carbondale: Southern Illinois University Press, 1986), 139-183.

47. Imposing a causal impact on the final vote is always a tricky process. In this instance, Wirthlin admitted that the "Morning in America" commercials may have created an overall climate in which voters were predisposed to move toward supporting Reagan's reelection. Nonetheless, research employing focus groups demonstrated that two other ads worked very efficiently in communicating the key bits of information associated with a vote shift for Reagan. Richard Wirthlin, in-class presentation at The Graduate School of Political Management, December 12, 1991.

48. Richard Wirthlin, telephone interview, April 27, 1992.

49. Kathleen Hall Jamieson, *Eloquence in an Electronic Age: The Transformation of Political Speechmaking* (New York: Oxford University Press, 1988), 240.

50. For a recent study of citizen attitudes toward politics, see Richard Harwood, "Citizens and Politics: A View from Main Street" (Dayton, Ohio: Kettering Foundation, 1991).

51. David H. Sawyer, "Confessions of a Political Consultant," an address to the University of Virginia Law School (Charlottesville, Va.: October 1991); and Kamber, "Trivial Pursuit."

52. I am alluding here to the problem of defining what we mean by negative ads; for a more formal discussion, see Johnson-Cartee and Copeland, *Negative Political Advertising,* 3-62.

53. Remarks of Malcolm MacDougall, an advertising executive who has some political experience, as quoted by Paul Taylor, *See How They Run:*

Electing the President in an Age of Mediaocracy (New York: Knopf, 1990), 215.

54. Many of these charges were voiced in a series of articles following the 1988 presidential elections written by reporters for the *Washington Post.* They reported numerous instances of politicians complaining that their consultants gave them no choice but to attack their opponents with negative ads. See Paul Taylor, "Consultants: Winning on the Attack," January 17, 1989, A1 and A14; Lloyd Grove, "How Experts Fueled the Race with Vitriol," January 18, 1989, A14; and David S. Broder, "Politicians, Advisers Agonize Over Campaigns' Character," January 19, 1989, A1 and A22.

55. Technically the Willie Horton ad was not produced by the Bush campaign, but was an independent expenditure by a political action committee. The Roger Ailes agency did produce an ad featuring a revolving prison door, which attacked Dukakis for the Massachusetts furlough program. See Hume, *Restoring the Bond,* 94-95.

56. Garramone et al., "Effects of Negative Political Advertising on the Political Process," 308.

57. Quoted by David Sawyer, "Confessions of a Political Consultant," 7.

58. Both Ellen Hume, "Restoring the Bond," and Victor Kamber, "Trivial Pursuit," start by critiquing ads but end by examining the importance of TV news and television in general.

59. See Kiko Addato, "The Incredible Shrinking Soundbite" (Cambridge, Mass.: Joan Shorenstein Barone Center, John F. Kennedy School of Government, Research Paper Number 2, June 1990).

60. An effort to demand that stations sell time for sixty-second commercials would probably succeed if any campaign had the determination to push the matter. Court decisions have generally sided with the right of the candidates to define what is "reasonable access," although winning in the courts is usually a pyrrhic victory achieved long after the election is settled. See *CBS v. FCC,* 101, S.Ct. 2813 (1981).

61. The conflicts between campaigns and broadcasters over access are often brutal, born of an incompatibility of standard operating procedures. Thomas Edmonds, presentation at The Graduate School of Political Management, June 1991.

62. Some consultants have even turned this argument into a justification (or explanation) for the rise of negative commercials. Hard-hitting attack ads can provide conflict and stark drama that will hold an audience that is otherwise prone to switching channels.

63. Popkin, *The Reasoning Voter,* 44-72.

64. James Fishkin, *Democracy and Deliberation: New Directions for Democratic Reform* (New Haven, Conn.: Yale University Press, 1991). See also Daniel Yankelovich, *Coming to Public Judgment* (Syracuse, N.Y.: Syracuse University Press, 1991); and Benjamin Barber, *Strong Democracy: Participatory Politics for a New Age* (Berkeley, Calif.: University of California Press, 1984).

65. David H. Sawyer, "Confessions of a Political Consultant," address to the University of Virginia Law School (Charlottesville, Va.: October 1991).

66. See Alexander and Bauer, *Financing the 1988 Elections,* 12. In an earlier study of escalating campaign costs, Alexander Heard noted that costs increased only gradually over much of this century. The sharp increases began in the 1950s just about the time that candidates turned to television advertisements. See Alexander Heard, *Elections in a Democracy* (Chapel Hill, N.C.: University of North Carolina Press, 1960).

67. Money in politics is a vast topic, only partially related to broadcast television commercials. See Elizabeth Drew, *Politics and Money: The New Road to Corruption* (New York: Macmillan, 1983); Brooks Jackson, *Honest Graft: Big Money and the American Political Process* (New York: Knopf, 1988); and David Magleby and Candice Nelson, *The Money Chase: Congressional Campaign Finance Reform* (Washington, D.C.: Brookings Institution, 1990).

68. Steven Bates, "Political Advertising Regulation: An Unconstitutional Menace?", *Cato Institute Policy Analysis* (Washington, D.C.: Cato Institute, September 22, 1988), 4.

69. See the report of the Harvard Study Group on Campaign Finance to the House Administration Committee, "An Analysis of the Impact of the Federal Election Campaign Act, 1972-1978." Committee on House Administration, U.S. House of Representatives (Washington, D.C.: G.P.O., October 1979).

70. See Alexander and Bauer, *Financing the 1988 Elections,* 12, for an estimate of soft-money accounts and their impact. Most of these complaints directed toward the soft money available for party-building activities derive from two other facts. First, in many cases, state laws are less restrictive as to the amounts that individuals can contribute. Second, the public reporting of these contributions is inadequate, so that public disclosure has been undercut.

71. Though the structure and conclusions of the following section are my own, I have benefited greatly from the extensive analysis conducted by Thomas H. Neale in "Negative Campaigning in National Politics: An Overview" (Washington, D.C., Congressional Research Service, Library of Congress, 91-775 GOV, September 18, 1991).

72. There have been some attempts to restrain campaign abuses on the state level as well; for a review, see James A. Albert, "The Remedies Available to Candidates Who Are Defamed by Television or Radio Commercials of Opponents," *Vermont Law Review* 5 (Spring 1986).

73. A recent detailed analysis uses a different categorization of legislative efforts. See Thomas H. Neale in "Negative Campaigning in National Politics: An Overview."

74. A Media General/Associated Press survey conducted September 8-15, 1986. Media General Inc., Richmond, Va. Cited in Neale, "Negative Campaigning in National Politics," 31. See also the survey released by the Times Mirror Center for the People and the Press reported in *The*

Polling Report, (December 17, 1990, vol. 6).

75. See the testimony of Curtis B. Gans, executive director of the Committee for the Study of the American Electorate, before the Senate Committee on Commerce, Science and Transportation. "The Clean Campaign Act of 1985: Hearings on S. 1310, September 10 and October 8, 1985," 99th Congress, 1st Session (Washington, D.C.: U.S. Government Printing Office, 1986), 29-30. John Lindsay, the former Mayor of New York City, also proposed that political TV ads be abolished at an Aspen Institute Conference, Wye Plantation, Maryland, June 1985.

76. For details on all the legislative proposals summarized here, see Neale, "Negative Campaigning in National Politics," 32-43.

77. Jack Winsbro, "Misrepresentation in Political Advertising: The Role of Legal Sanctions," *Emory Law Journal* 35 (Summer 1987): 913; quoted from Neale, 35.

78. Neale, "Negative Campaigning in National Politics," 36.

79. Bates, "Political Advertising Regulation."

80. *The New York Times v. Sullivan,* 376 U.S. 254 (1964).

81. Ellen Hume, "Restoring the Bond," 94-95.

82. The fact that journalists may develop a sympathy toward the candidates they cover was first pointed out by Timothy Crouse in *Boys on the Bus* (New York: Ballantine, 1973); Christopher Arterton, *Media Politics: The News Strategies of Presidential Campaigns* (Lexington, Mass.: Lexington Books, 1984); see also Judith Lichtenberg, "The Journalist's Duty to Betray," National Press Club Record, July 20, 1989.

83. *Washington Post,* January 19, 1989, A22.

84. See, for example, two cases studies on ethics in politics written by Gregory Lebel, "The Campaign Watchdog: Dade County's Fair Campaign Practices Committee and the Race for the 'American Seat' " (Washington, D.C.: Graduate School of Political Management, case number 90-1, 1990); and "CONDUCT and the 1987 Chicago Mayor's Race" (Washington, D.C.: Graduate School of Political Management, case forthcoming).

85. Kamber, "Trivial Pursuit," 36.

86. *Vanasco v. Schwartz,* 401 F. Supp. 87 (DCNY, 1975). See also Leslie Tucker and David Heller, "Putting Ethics into Practice," *Campaigns and Elections* (March/April 1987): 42-46. The limiting effect of the First Amendment remains to be fully tested. Recently, for example, a Minnesota statute against lying in campaigns was invoked when a candidate was indicted on criminal charges for false attacks on his opponent. The case was thrown out for lack of evidence. The Minnesota law has been on the books for eighty years. See Tom Smalec, "A Law against Lying," *Campaigns and Elections* (August 1989): 49.

87. See "Political Consultants Try to Put House in Order, But Critics Dismiss Move as Public Relations," *Campaign Practices* 16, February 6, 1989.

88. The history of these efforts goes back at least as far as Ray Bliss's

injunctions in the 1960s that Republicans should "not speak ill of fellow Republicans." Most recently, in the middle of the 1992 nomination campaign, Democratic Chair Ron Brown publicly criticized candidate Jerry Brown for his rhetorical excesses aimed at Bill Clinton; see Gwen Ifill, "Campaign Attacks by Brown Assailed by His Party Chief," *New York Times,* March 27, 1992, A1.

89. Valerie Richardson, "Atwater, Brown Agree to Discuss Dirty Campaigning, Redistricting," *Washington Times,* November 6, 1989, A5.
90. Neale, "Negative Campaigning in National Politics," 46.
91. Strategically the candidate running behind usually reaches for sharp critical ads first in an effort to even the race by tarnishing the luster of his or her opponent. More recently, even congressional incumbents who start with a substantial advantage have, in some cases, launched an early massive attack in an attempt to keep their challenger from getting started.
92. See the papers produced for the "Conference on Professional Responsibility and Ethics in the Political Process," (Washington, D.C.: American Association for Political Consultants, March 24, 1991); E. J. Dionne, "Political Consultants Debate Professional Ethics," *Washington Post,* March 25, 1991, A4; Randall Rothenberg, "Image Makers Look at Themselves," *New York Times,* March 25, 1991, A14; and Paul O. Wilson, "Establishing a Code of Professional Responsibility for Political Consultants" (Alexandria, Va.: Wilson Communications, mimeo, May, 1991).
93. Robert K. Fullinwider, in F. Christopher Arterton and Robert Fullinwider, "Ethical Problems in American Politics" (New York: Graduate School of Political Management, a concept paper accompanying proposal for establishment of a program in ethics in politics, 1992). Fullinwider cites Brad O'Leary, president of the American Association of Political Consultants in January 1989, who once opined that "[n]o . . . member sees any validity in what is being said" about the contribution of consultants to the low quality of campaign rhetoric. "I just don't think an unfair ad can run today," he said, "without the press killing the campaign that runs it." *Washington Post,* January 19, 1989, A1.
94. Undoubtedly the skill demonstrated by Ailes brought additional business, but he felt himself to be under considerable criticism. It is possible that he did not undertake the ad campaign for Bush's reelection bid in 1992 in part because of the controversy and criticism, although such decisions depend on many factors.

4. OPEN SEASON: HOW THE NEWS MEDIA COVER PRESIDENTIAL CAMPAIGNS IN THE AGE OF ATTACK JOURNALISM

Larry J. Sabato

Larry Sabato, in examining recent developments in campaign coverage by the news media, provides support for John Aldrich's contention that a candidate's message is irrelevant to the campaign and to the outcome of the election. Sabato argues that structural changes in media markets and a loosening of libel law in 1964 led political news coverage to become more entertainment-oriented and sensationalist. Reporters cover eye-catching, amusing events rather than policy speeches and position papers, and editors and producers are scarcely reticent about highlighting scandal, rumor, and innuendo to the detriment of substantive campaign news.

Moreover, Sabato argues, the competitive market for viewers forces reporters to go into a "feeding frenzy" over the kinds of stories that boost ratings. Thus, because television has nationalized the news market and the immediacy of news coverage, stories can develop their own momentum and "attractive" ones are picked up by the networks and newspapers. For example, campaign professional Mike Deaver re-counted a conversation he had with a reporter concerning a rumor about George Bush: "And this person said to me, 'Well, if it had been in the Washington Post, *we'd have had to go with it.' I said, 'Wait a minute, now. You're telling me that you would have told 80 million people about this based on one newspaper story?' [The person responded:] 'Oh, sure. We'd have to.' "*

Deaver went on to say, "the thing about television that's scary to me, is that they have got 22 minutes or 18 minutes for the evening news, and they are going to cram all the stuff in there. And they will say, 'The Miami Herald *today said that so-and-so is caught with his pants down, and the weather tomorrow—stay tuned for the sports,' and so forth. And that's it. And a guy has been ruined."*

According to Sabato, virtually everybody involved in political campaigns—candidates, reporters, editors, and producers—would pre-fer to focus more on candidates' issue positions and less on their personal lives. But, so long as the media want to survive in the business of reporting on politics and so long as voters prefer to hear about scandal, the incentive is to present personalities and catchy soundbites. —Ed.

The press has played many roles and assumed many postures in U.S. history. Moreover, like every other American institution, its power, promise, and performance have varied greatly from era-to era. In the modern age—from World War II to the present—the breadth, depth, and influence of the media have expanded enormously, marked by three successive phases in the development of political journalism.

From 1941 to 1966, journalists engaged in what I would call lapdog journalism—reporting that served and reinforced the political establishment. Mainstream journalists rarely challenged prevailing orthodoxy, accepted at face value much of what those in power told them, and protected politicians by revealing little about their nonofficial lives, even when private vices affected public performance. Wartime necessities encouraged the lapdog mentality that had become well established in Franklin Roosevelt's earlier administrations. However, lapdog journalism perhaps reached its zenith under another Democratic president, John F. Kennedy.

The period from about 1966 to 1974 was one of watchdog journalism, when reporters scrutinized the behavior of political elites by undertaking independent investigations of their statements. The Vietnam War and the Watergate scandal stimulated this kind of journalism. In the latter part of this period, the private life of officials began to be discussed, but usually only in the context of public performance.

Since about 1974, political reporters have engaged in what I would call junkyard-dog journalism—reporting that is often harsh, aggressive, and intrusive. The news media, both print and broadcast, have sometimes resembled piranhas or sharks in a feeding frenzy. Mere gossip can reach print, and every aspect of a person's private life potentially becomes fair game for scrutiny as a new, almost anything-goes philosophy takes hold.

Of course, the above are only generalizations, describing the prevailing mainstream attitude of journalism in each period: there are many exceptions in each time frame, and the periods overlap in some respects. Nonetheless, these three phases are useful barometers of the behavioral changes that have occurred in recent years.

In the following pages I will examine some of the reasons for these developments with respect to presidential elections, concentrating on structural changes in the news media, heightened competition in the press, the phenomenon of pack journalism, the influence of the Watergate scandal, and other factors. Two case studies from the 1988 presidential election, involving Michael Dukakis and Dan Quayle, will be cited to illustrate how changes in press coverage have affected presidential campaigns. Finally, the consequences of these changes for the candidates, the voters, and the political system will be assessed.

Reasons for the Changes

Technology and Growing Intensity

Conditions are always ripe for spawning a frenzy in the brave new world of the omnipresent press. Advances in media technology have revolutionized campaign coverage. Hand-held miniature cameras (minicams) and satellite broadcasting have enabled television to go anywhere, anytime, with ease. Instantaneous transmission (by broadcast and fax) to all corners of the country has dramatically increased the velocity of campaign developments today, accelerating events to their conclusion at breakneck speed. Gary Hart, for example, went from front-runner to ex-candidate in less than a week in May 1987. Continuous public affairs programming, such as C-SPAN and CNN, helps put more of a politician's utterances on the record. C-SPAN, CNN, and satellite broadcasting capability also contribute to the phenomenon called "the news cycle without end," which creates a voracious news appetite demanding to be fed constantly. This increases the pressure to include marginal bits of information and gossip and produces novel if distorting "angles" on the same news to differentiate one report from another. The extraordinary number of local stations covering national politics—up to several hundred at major political events—creates an echo chamber, with seemingly endless repetition of essentially the same news story. This local contingent also swells the press corps traveling the campaign trail. In 1988, for instance, an estimated two thousand journalists of all stripes flooded the Iowa caucuses.[1] Reporters not infrequently outnumber the audience at meetings and whistlestops.[2]

Oddly enough, just as the coverage of presidential elections is expanding, the quality of that coverage has become very uneven and often deeply disappointing. Superficial "horse-race" reporting—a focus on who's ahead, who's behind, and who's gaining—has been a problem for many decades and is now the norm. Of seven thousand print news stories surveyed between Labor Day and election day 1988, 57 percent were horse-race items and only 10 percent concentrated on real policy issues.[3] On television network news shows during the 1988 primary campaign, two and a half times more horse-race pieces aired than did those devoted to policy issues.[4] Reporters love the horse race rather than policy for many different reasons. Most journalists are generalists, unfamiliar with the nuances and complexities of many issues and therefore unprepared to focus on them in news stories. Then, too, the horse race fits many reporters' rather cynical view of politics: elections are but a game, not a contest of the power of competing ideas. And horse-race reporting has no partisan bias, being far more objective and disinterested than the construction of an issue agenda for coverage. The length of modern campaigns also plays a part; reporters may be trying to stave off boredom. Tom Rosenstiel, media reporter of the *Los Angeles Times*, compares American elections not to a horse race but to a football game, where "we play about forty-seven quarters before any of the fans have come into the stadium. The players are on the field and the press is in the press box, and we have played for months before anybody cares or watches." [5] Such boredom leads the press from time to time to stop analyzing the game and concentrate on the individual players—the source of more than one feeding frenzy over the years. This gamesmanship coverage of politics discourages sober discussion of policy choices while fostering personal conflict, controversy, and confrontation.

Competitive Pressures in the Media

The media, of course, have always been in the business of selling themselves. And sex and scandal do indeed sell well—a great deal better than dispassionate discussion of policy issues. But while this has always been the case, there are now added inducements for coverage of scandal during presidential elections. First and foremost, corporate managers of broadcast and print media are closely attuned to the contribution each outlet makes to the overall profit or loss of the company. Ratings and circulation grow more important as profits become more difficult to generate in an increasingly competitive industry. Executives are not unaware of the experiences of newspapers such as the *Washington Times*, whose screaming front-page headlines about Rep. Barney Frank and an unrelated homosexual "call-boy" scandal boosted newsstand sales by 25 percent in the summer of 1989. Nor are individual reporters, in the age of "star journalism," oblivious to the rewards that may await them at scandal's end. As syndicated columnist Robert Novak explains: "There are very high rewards in both money and in fame, much more than when I started. A lot of very bright young men and women want the rewards, and they know damn well that carefully covering Fairfax County [Virginia] is not going to do it for them." [6]

The competitive philosophy governing this new generation of go-getters can be summed up by their fevered adherence to journalism's two ultimate imperatives: "Don't get beaten to a story by another media outlet," and "If we don't break this, someone else will." Both axioms, especially the second, serve to encourage bad judgment and premature decisions to publish or air stories. During Watergate, for example, according to Gerald Warren, President Richard Nixon's deputy press secretary, "The competitive urge in Washington . . . was so great that [the press] would go after anything, and so there were a number of cases where rumors were put on the wire [first] and then checked out."

Competitive pressures also propel the efforts of each newspaper and network to make a contribution, no matter how small and insignificant, to the unfolding frenzy. Each new fact merits a new story, complete with a recapitulation of the entire saga. This press pile-on by means of minor "wiggle" disclosures, an ordeal akin to water torture (drip, drip, drip, day after day after day), is well known to many presidential press victims. In 1987, for example, every newly uncovered, unattributed quotation borrowed by a Biden speechwriter became a news item. Four years earlier, during Geraldine Ferraro's time of trial over family finances, the exasperated vice-presidential candidate sighed, "Every day there is another story about another building." In sum, wiggle disclosures extend and intensify a feeding frenzy while stretching thin the available news and reiterating it excessively.

Pack Journalism

Oddly enough, these intense competitive pressures cause reporters (as well as editors and producers) to move forward together in essentially the same story direction rather than on different tracks. This condition, called "pack journalism," prompted former U.S. Senator Eugene McCarthy to liken it to blackbirds on a telephone wire: when one moves to another wire they all move. A less colorful but operationally accurate description of pack journalism has been offered by George Skelton of the *Los Angeles Times:*

> When you're covering the White House and twenty reporters are listening to the President's speech, someone comes up to the most senior reporter and says, "What's the lead?" And several reporters say, "Yeah, that makes sense to me." And pretty soon . . . eighteen out of the twenty reporters [are] using the same lead. That's pack journalism.

Partly for the same reason, the free market of journalism often leads to uniformity rather than diversity. "Journalists want to do something that their peers think is meaningful," notes Jim Gannon, Washington bureau chief of the *Detroit News.* "Peer pressure matters in this big fraternity, and so you tend to turn to the sexy topic of the hour." [7]

Most frequently a major newspaper breaks the initial "big scandal story" and points the herd in a certain direction. But just as often today, a widely watched television broadcast—the evening news or a Sunday commentary show—will give the first indication of the conventional wisdom that will develop in reaction to the print story. "This Week with David Brinkley," the program on ABC, did the honors for Gary Hart after the *Miami Herald*'s same-day publication of the Donna Rice affair, with panelist Sam Donaldson, among others, suggesting that Hart's candidacy was nearing its end.

Some have seen a change in pack journalism that offers hope for reduced groupthink.[8] For example, networks and newspapers for the most part no longer assign correspondents to travel with a single candidate for the duration of the campaign, potentially lessening the cocoon-like effect of a static entourage. Yet there is little evidence that pack journalism has abated, and in some ways its grip on the press corps, especially television reporters, may be strengthening. Ted Koppel, of ABC News and host of "Nightline," remembers that nonconformist journalism was more likely when production expectations were lower for a TV newsman:

> When I was a foreign correspondent twenty years ago, I worked just for the evening news. Now a foreign correspondent for ABC has got to worry about the demands of "World News Tonight," "Good Morning America," "Nightline," "Prime Time Live," "David Brinkley," and "20/20." What does that mean? More and more pressure to get the pieces on the air quickly. Less and less time to go out and do the actual reporting. So you end up going with the flow. It takes much more time to dig into a story on your own than it does simply to match what you know your competition is doing.[9]

Whether on the rise or not, the unfortunate effects of pack journalism—conformity and formulaic reporting—are apparent to both news reporters and news consumers. Innovation is discouraged, and the checks and balances supposedly provided by competition evaporate. Press energies are devoted to finding mere variations on a theme (new angles and wiggle disclosures), while a mob psychology catches hold that allows little mercy for the victim.

Watergate's Watermark on the Media

The Watergate scandal had the most profound impact of any modern event on the manner and substance of press conduct. In many respects it launched the media's open season on politicians in a chain reaction that today allows scrutiny of even the most private sanctums of public officials' lives. Moreover, combined with Vietnam and the civil rights movement, Watergate shifted the orientation of journalism away from mere description—providing an accurate account of happenings—and toward prescription—helping to set the campaign's (and society's) agendas by focusing attention on the candidates' shortcomings as well as certain social problems.

During this time a new breed and a new generation of reporters were attracted to journalism, particularly its investigative arm. As a group they were idealistic, though aggressively mistrustful of authority, and shared a contempt for politics as usual. Critics called them do-gooders and purists who wanted the world to stand at moral attention for them.

Many who found journalism newly attractive in the wake of Watergate were, of course, not completely altruistic. The ambitious saw the happy fate of the *Washington Post*'s young Watergate sleuths Bob Woodward and Carl Bernstein, who gained fame and fortune, not to mention big-screen portrayals by Robert Redford and Dustin Hoffman in the movie *All the President's Men*. As *U.S. News & World Report*'s Steven Roberts sees it:

> A lot of reporters run around this town dreaming of the day that Dustin Hoffman and Robert Redford are going to play them in the movies. That movie had more effect on the self-image of young journalists than anything else. Robert Redford playing a journalist? It lends an air of glamour and excitement that acts as a magnet drawing young reporters to investigative reporting.[10]

The young were attracted not just to journalism but to a particular kind of journalism. The role models were not respected, established reporters but two unknowns who refused to play by the rules their seniors had accepted. "Youngsters learned that deductive techniques, all guesswork, and lots of unattributed information [were] the royal road to fame, even if it wasn't being terribly responsible," says Robert Novak.[11] After all, adds columnist and television commentator Mark Shields, "Robert Redford didn't play Walter Lippman and Dustin Hoffman didn't play Joseph Kraft."[12]

The Character Issue Takes Hold

A clear consequence of Watergate and other recent historical events has been the media's increasing emphasis on the character of presidential candidates. As journalists assessed the three tragic but exceptionally capable figures who held the presidency from 1960 to 1974, they saw that the failures of Kennedy, Johnson, and Nixon were not those of intellect but of ethos.

The issue of character has always been present in American

politics—not for his policy positions was George Washington made our first president—but rarely, if ever, has character been such a pivotal concern in presidential elections, both primary and general, as it has since 1976. The 1976 Carter campaign was characterized by considerable moral posturing; Edward Kennedy's 1980 candidacy was in part destroyed by lingering character questions; Walter Mondale finally overcame Gary Hart's 1984 challenge in the Democratic primaries by using character as a battering ram; and 1988 witnessed such a forceful explosion of concern about character that several candidates were eliminated and others badly scarred by it.

Whatever the precise historical origins of the character trend in reporting, it is undergirded by certain assumptions—some valid, others dubious. First and most important of all, *the press correctly perceives that it has mainly replaced the political parties as the "screening committee" that winnows the field of candidates and filters out the weaker or more unlucky contenders.* Second, many reporters, again correctly, recognize the mistakes made under the rules of lapdog journalism and see the need to tell people about candidate foibles that affect public performance. Third, the press assumes that it is giving the public what it wants and expects, more or less. Television is the primary factor here, having served not only as handmaiden and perhaps mother to the age of personality politics but also conditioning its audience to think about the private lives of "the rich and famous."

Less convincing, however, are a number of other assumptions about elections and the character issue made by the press. Some journalists insist upon their obligation to reveal everything of significance discovered about a candidate's private habits; to do otherwise, they say, is antidemocratic and elitist.[13] Such arguments ignore the press's professional obligation to exercise reasonable judgment about what is fit to be printed or aired as well as what is most important for a busy and inattentive public to absorb. Other reporters claim that character matters so much because policy matters so little, that the issues change frequently and the pollsters and consultants determine the candidates' policy stands anyway.

Perhaps most troubling is the almost universally accepted belief that private conduct affects the course of public action. Unquestionably, private behavior can have public consequences. However, it is far from certain that private vice inevitably leads to corrupt, immoral leadership or that private virtue produces public good. Indeed, the argument can be made that many lives run on two separate tracks (one public, one private) that should be judged independently. In any event, a focus on character becomes not an attempt to construct the mosaic of qualities that make up an individual but rather a strained effort to find a sometimes manufactured pattern of errors or shortcomings that will automatically disqualify a candidate.

To be sure, the Hart character frenzy was worrying to many journalists. Lines had been crossed, and methods used, that gave pause. Steven Roberts of *U.S. News & World Report* "object[s] in many ways" to the Hart story:

> I just don't like the idea of reporters sitting in the bushes watching someone's bedroom, which is basically what the *Miami Herald* did. But in hindsight I think it was very important that the material came out. I didn't like the means, but I liked the end, because the end clearly told us something critically important about Hart's ability to be president. It told us something profoundly important about his sense of judgment, something directly relevant to his capacity to be president. So in the end, the process worked, but I was always very uncomfortable with the way it worked.[14]

Roberts's doubts were widely shared, but so was his conclusion. Bad means had yielded a good end—Hart's demise as a serious presidential contender. In the process, however, those bad means may have been sanctified and, without doubt, the character issue had become even more firmly rooted in journalistic practice.

Press Frustration

Not surprisingly, politicians react rather badly to the treatment they receive from the modern press. Convinced that the media have but one conspiratorial goal—to hurt or destroy them—the pols respond by restricting journalists' access, except under highly controlled situations. Kept at arm's length and out of the candidate's way, reporters have the sense of being enclosed behind trick mirrors: they can see and hear the candidate, but not vice versa. Their natural, human frustrations grow throughout the grueling months on the road, augmented by many other elements, including a campaign's secrecy, deceptions, and selective leaks to rival newsmen, as well as the well-developed egos of candidates and their staffs. Despite being denied access, the press is expected to provide visibility for the candidate, to retail his or her bromides. Broadcast journalists especially seem trapped by their need for good video and punchy soundbites and with regret find themselves falling into the snares set by the campaign consultants—airing verbatim the manufactured message and photoclip of the day. The press's enforced isolation and the programmed nature of its assignments produce boredom as well as disgruntlement, yet the professionalism of the better journalists will not permit them to let their personal discontent show in the reports they file.

These conditions inevitably cause reporters to strike back at the first opportunity. Whether it is emphasizing a candidate gaffe, airing an unconfirmed rumor, or publicizing a revelation about the candidate's personal life, the press uses a frenzy to fight the stage managers, generate some excitement, and seize control of the campaign agenda. Media emotions have been so bottled and compressed that even the smallest deviation from the campaign's prepared script is trumpeted as a major development.

This scenario aptly describes the 1988 presidential campaign, especially the general election. George Bush and his experienced aides were exceptionally successful in keeping the press at bay. (The expertly media-managed Reagan presidency served as training ground and role model.) Because Bush was so inaccessible and ran a relatively mistake-free post Labor-Day campaign, his message was rarely "stepped on," as the terminology goes—that is, the press never got control of the agenda after the Quayle frenzy subsided. Nonetheless, reporters' frustration was evident in the intense coverage showered on Bush's main verbal gaffe—his innocuous misstating of the date of Pearl Harbor Day. Michael Dukakis's organization proved slow to grasp and match Bush's strategy. After granting easy access for months, and having the press step all over his message as a consequence, Dukakis began to emulate Bush and became less available to the press in October.

Bias

Does press frustration, among other factors, ever result in uneven treatment of presidential candidates, a tilt to one side or the other, further helping to foster attack journalism? In other words, are the news media biased? One of the enduring questions of journalism, its answer is simple and unavoidable: of course they are. Journalists are fallible human beings who inevitably have values, preferences, and attitudes galore—some conscious and others subconscious—all reflected at one time or another in the subjects or slants selected for coverage. To revise and extend the famous comment of Iran-Contra defendant Oliver North's attorney Brendan Sullivan, reporters are not potted plants.

The vital question, then, is in what way is the press biased? First of all, on an ideological level, the press is liberal. "There is a liberal tilt among the press that we refuse to recognize, and it is pointless not to acknowledge it because it is obvious," asserts Michael Barone, a senior writer for *U.S. News & World Report*.[15] The relatively small group of journalists (not much over one hundred thousand, compared to four million teachers in the United States) is drawn heavily from the ranks of highly educated social and political liberals, as confirmed by a number of studies, some conducted by the media.[16] Journalists are substantially Democratic in party affiliation and voting habits, progressive and antiestablishment in political orientation, and well to the left of the general public on most economic, foreign policy, and especially social issues (such as abortion, affirmative action, gay rights, and gun control). Second, many dozens of the most influential reporters and executives entered or reentered journalism after stints of partisan involvement in campaigns or government, and a substantial majority worked for the Democrats.[17] Third, this liberal press bias does indeed show up frequently on screen and in print.

A study of reporting on the abortion issue, for example, revealed a clear slant to the pro-choice side on network television news, matching in many ways the reporters' own abortion-rights views.[18] The *Washington Post* gave extensive coverage to a 1989 pro-choice rally that attracted 200,000 (a dozen stories, leading the front page), while nearly ignoring a pro-life rally that attracted 200,000 (two stories in the Metro section).[19] *Post* reporters have also repeatedly been observed cheering speeches at pro-choice rallies.[20] Additionally, the press falls to the left on the topics they choose to cover.[21] In the latter half of the 1980s, for instance, television gave enormous attention to the homelessness issue; in the process, some experts in the field believe the broadcasters both exaggerated the problem and grossly overstated the role of unemployment in creating the homeless.[22]

Conservative politicians do not need a Gallup survey to convince them of press bias against their ilk. For decades, right-of-center Republican leaders have railed against the "liberal media." When President Dwight Eisenhower attacked the "sensation-seeking" press at the 1964 GOP convention that nominated Barry Goldwater for president, delegates raged against the media in attendance, cursing and shaking their fists at them.[23] Conservative journalists such as Republican activist Patrick J. Buchanan have built careers assailing their liberal brethren. Wrote Buchanan in one column, "America is bitter because the front pages of the prestige press and the major network 'news' shows are saturated with liberal bigotry, the practitioners of which are either too blind to see it or too dishonest to concede it." [24]

The 1984 and 1988 campaigns stoked conservative fires. Despite the controversy over Geraldine Ferraro's finances, the Mondale-Ferraro team received far kinder, more positive coverage than did the Reagan-Bush team, according to a study by political scientists Michael J. Robinson and Maura Clancey.[25] Four years later George Bush again received a highly negative press, although Michael Dukakis fared not much (if at all) better.[26] Most rankling to conservatives, however, were these two findings: the most liberal candidate for president in

1988, Jesse Jackson, garnered the best press, while the most conservative of the four national party nominees, Dan Quayle, received the worst.[27]

More essential to understanding press bias are the nonideological factors. Owing to competition and the reward structure of journalism, the deepest bias most journalists have is the desire to get to the bottom of a good campaign story. Indeed, pack journalism is more of a factor than bias in prompting all media outlets to focus on the same developing "good story" and encouraging them to adopt the same slant.

A related nonideological bias is the effort to create a horse race where none exists.[28] News people whose lives revolve around the current political scene naturally want to add spice and drama, minimize the boredom, and increase their audience. Runaway elections such as in 1984 inevitably find the press welcoming a new face (Hart)[29] or trying to poke holes in the campaign of the heavy favorite (Reagan).

In their quest to avoid bias, reporters also frequently seize on nonideological offenses such as gaffes, ethical violations, and campaign finance problems. These "objective" items are intrinsically free from partisan taint and can be pursued with the relish denied the press on "hot button," party-polarizing issues. Finally, other human, not just partisan, biases are at work. Whether the press likes or dislikes a candidate is often vital. Former Arizona Governor Bruce Babbitt, for instance, was a press favorite and enjoyed favorable coverage both as governor and presidential candidate in 1988. Conversely, Richard Nixon, Jimmy Carter, and Gary Hart were roundly disliked by many reporters and were given much unfavorable coverage.

In sum, then, press bias of all kinds—partisan, agenda setting, and nonideological—has influenced the development of junkyard-dog journalism in covering presidents and presidential candidates. But ideological bias is not the be-all and end-all that critics on both the right and left often insist it is. Press tilt has a marginal effect, no more, no less.

Two Cases of Attack Journalism in the 1988 Presidential Election: Dukakis and Quayle

Michael Dukakis's 1988 mental-health controversy is one of the most despicable episodes in recent American politics. The corrosive rumor that the Democratic presidential nominee had undergone psychiatric treatment for severe depression began to circulate in earnest at the July 1988 national party convention. The agents of the rumormongering were "LaRouchies," adherents of the extremist cult headed by Lyndon LaRouche, who claims, among other loony absurdities, that Queen Elizabeth II is part of the international drug cartel.[30]

Shortly after the Democratic convention, the Bush campaign—

with its candidate trailing substantially in the polls—began a covert operation to build on the foundation laid by the LaRouchies. As first reported by columnists Rowland Evans and Robert Novak,[31] Bush manager Lee Atwater's lieutenants asked outside Republican operatives and political consultants to call their reporter contacts about the matter. These experienced strategists knew exactly the right approach in order not to leave fingerprints, explains Steve Roberts of *U.S. News & World Report:*

> They asked us, "Gee, have you heard anything about Dukakis's treatment? Is it true?" They're spreading the rumor, but it sounds innocent enough: they're just suggesting that you look into it, and maybe giving you a valuable tip as well.[32]

Many newspapers, including the *Baltimore Sun* and the *Washington Post,* at first refused to run any mention of the Dukakis rumor since it could not be substantiated.[33] But on August 3 an incident occurred that made it impossible, in their view, not to cover the rumor. During a White House press conference a correspondent for *Executive Intelligence Review,* a LaRouche organization magazine, asked Reagan if he thought Dukakis should make his medical records public. A jovial Reagan replied, "Look, I'm not going to pick on an invalid." Reagan half apologized a few hours later ("I was just trying to be funny and it didn't work"), but his weak attempt at humor propelled into the headlines a rumor that had been only simmering on the edge of public consciousness.

Whether spontaneous or planned, there is little doubt that "Reagan and the Bush people weren't a bit sorry once it happened," as CNN's Frank Sesno asserts.[34] The Bush camp immediately tried to capitalize on and prolong the controversy by releasing a report from the White House doctor describing their nominee's health in glowing terms.[35] But this was a sideshow compared with the rumor itself. The mental-health controversy yanked the Dukakis effort off track and forced the candidate and then his doctor to hold their own press conference on the subject, attracting still more public attention to a completely phony allegation. False though it was, the charge nonetheless disturbed many Americans, raising serious doubts about a candidate who was still relatively unknown to many of them. "It burst our bubble at a critical time and cost us half our fourteen-point [poll] lead," claims the Dukakis staff's senior adviser, Kirk O'Donnell. "It was one of the election's turning points; the whole affair seemed to affect Dukakis profoundly, and he never again had the same buoyant, enthusiastic approach to the campaign."[36]

As is usually the case, the candidate unnecessarily complicated his own situation. Until events forced his hand, Dukakis stubbornly refused to release his medical records or an adequate summary of them despite advance warning that the mental-health issue might be raised. But the press can by no means be exonerated. While focusing on the relatively innocent casualty, most journalists gave light treatment to the perpetrators. In retrospect, several news people said they regretted not devoting more attention to the LaRouche role in spreading the rumor, given his followers' well-deserved reputation as "dirty tricksters." [37]

Overall, one of the most important lessons of the Dukakis mental-health episode is that caution must be exercised in reporting on presidential campaign rumors. "The media are really liable for criticism when we get stampeded by competitive instincts into publishing or airing stories that shouldn't be on the record," says National Public Radio's Nina Totenberg. "We were stampeded on the Dukakis story, and we should never have let it happen." [38]

The perils of vice-presidential candidate Dan Quayle became perhaps the most riveting and certainly the most excessive feature of 1988's general election. For nearly three weeks, coverage of the presidential campaign became mainly coverage of Quayle. Most major newspapers assigned an extraordinary number of reporters to the story (up to two dozen), and the national networks devoted from two-thirds to more than four-fifths of their total evening-news campaign minutes to Quayle.[39] Combined with the juicy material being investigated, this bumper crop of journalists and stories produced, in the words of a top Bush/Quayle campaign official, "the most blatant example of political vivisection that I've ever seen on any individual at any time; it really surpassed a feeding frenzy and became almost a religious experience for many reporters." Balance in coverage, always in short supply, was almost absent. First one controversy and then another about Quayle's early life mesmerized the press, while little effort was made to examine the most relevant parts of his record, such as his congressional career.

It was the big-ticket items about Quayle—his National Guard service, the alleged love affair with Paula Parkinson, and his academic record—that attracted the most attention. At the convention, wild rumors flew, notably the false allegation that Quayle's family had paid fifty thousand dollars to gain him admission to the Guard. It was unquestionably legitimate for the press to raise the National Guard issue, although once the picture became clear—Quayle's family did pull strings, but not to an unconscionable degree—some journalists appeared unwilling to let it go. Far less legitimate was the press's resurrection of a counterfeit, dead-and-buried episode involving lobbyist Paula Parkinson. As soon as Quayle was selected for the vice-presidential nomination, television and print journalists began mentioning the 1980 sex-for-influence "scandal," despite the fact that Quayle had long ago been cleared of any wrongdoing and involvement with Parkinson. "When Quayle's name came up as a vice-presidential possibility, before his selection, the word passed among reporters that Bush couldn't choose Quayle because of his 'Paula problem,' " admitted one television newsman. "It was the loosest kind of sloppy association . . . as if nobody bothered to go back and refresh their memory about the facts of the case."

Some of the rumors about Quayle engulfing the press corps stretched even farther back into his past than did the womanizing gossip. Quayle's academic record was particularly fertile ground for rumormongers. By his own admission, the vice-presidential nominee had been a mediocre student, and the evidence produced during the campaign suggests that mediocre was a charitable description. At the time, however, a rumor swept through Quayle's alma mater, DePauw University, that he had been caught plagiarizing during his senior year. This rumor, which cited a specific teacher and class, was widely accepted as true and became part of the Quayle legend on campus.

Within a day of Quayle's selection as the vice-presidential nominee, the rumor had reached the New Orleans GOP convention hall. Hours after the convention was adjourned, the *Wall Street Journal* published a lengthy article on Quayle's problems, noting unsubstantiated "rumors" of a "cheating incident." [40] This story helped to push the plagiarism rumor high up on the list of must-do Quayle rumors, and soon the press hunt was on—for every DePauw academic who had ever taught Quayle, for fellow students to whom he might have confided his sin, even for a supposedly mysterious extant paper or bluebook in which Quayle's cheating was indelibly recorded for posterity.

As it happens, the plagiarism allegation against Quayle appears to have a logical explanation, and it was apparently first uncovered by the painstaking research of two *Wall Street Journal* reporters, Jill Abramson and James B. Stewart (the latter a graduate of DePauw, which fortuitously gave him a leg up on the competition). Abramson and Stewart managed to locate almost every DePauw student who had been a member of Quayle's fraternity, Delta Kappa Epsilon, during his undergraduate years. Approximately ten did remember a plagiarism incident from 1969 (Quayle's year of graduation), and the guilty student was in fact a golf-playing senior who was a political science major and a member of the fraternity—but *not* Quayle. The similarities were striking and the mix-up understandable after the passage of nearly twenty years. What was remarkable, however, was the fact that an undistinguished student such as Quayle would be so vividly remembered by the faculty. Abramson and Stewart also uncovered the reason for this, and even two decades after the fact their finding makes a political science professor blanch. Quayle was one of only two 1969 seniors to fail the political science comprehensive exam, a requirement for graduation. (He passed it on the second try.) Abramson's conclusion was reasonable: "Jim Stewart and I believed that people had confused Quayle's failure on the comprehensive exam with his ... fraternity brother's plagiarism, especially since both events ... occurred at the same time."[41] Unfortunately for Quayle, however (and also for the public), this explanation did not reach print, even though it might have provided a fair antidote to the earlier rumor-promoting article. Instead, the assumption that Quayle must have cheated his way through college solidified and led to other academically oriented rumors and questions, among them how a student with such a poor undergraduate record could gain admission to law school.

An observer reviewing the academic stories about Quayle is primarily struck by two elements. First, despite the windstorm of rumor that repeatedly swept over the press corps, there was much fine, solid reporting, with appropriate restraint shown about publishing rumors, except for the original *Journal* article mentioning plagiarism and some pieces about Quayle's law-school admission. Of equal note, however, was the overwhelming emphasis on his undergraduate performance. As any longtime teacher knows, students frequently commit youthful errors and indiscretions that do not necessarily indicate their potential or future development. Thus, once again, the question of balance is raised. How much emphasis should have been placed on, and precious resources devoted to, Quayle's life in his early twenties compared with his relatively ignored senatorial career in his thirties?

Consequences

Having examined some of the truths about feeding frenzies, we now turn to their consequences. Attack journalism has major repercussions on the institution that spawns it—the press—including how it operates, what the public thinks of it, and whether it helps or hurts the development of productive public discourse. The candidates and their campaigns are also obviously directly affected by the ways and means of frenzy coverage, in terms of which politicians win and lose and the manner of their running. The voters' view of politics—optimistic or pessimistic, idealistic or cynical—is partly a by-product of what they learn about the subject from the news media. Above all, the dozens of feeding frenzies in recent times have had substantial and cumulative effects on the American political system, not only determining the kinds of issues discussed in campaigns but also influencing the types of people attracted to the electoral arena.

One of the great ironies of contemporary journalism is that the effort to report more about candidates has resulted in the news media often learning less than ever before. Wise politicians today regard their every statement as being on the record, even if not used immediately—perhaps turning up the next time the news person writes a profile. Thus the pols are much more guarded around journalists than they used to be, much more careful to apply polish and project the proper image at all times. The dissolution of trust between the two groups has meant that "journalists are kept at an arm's length by fearful politicians, and to some degree the public's knowledge suffers because reporters have a less well-rounded view of these guys," says Jerry terHorst, Gerald Ford's first press secretary and former *Detroit News* reporter.[42] The results are easily seen in the way in which presidential elections are conducted. Ever since Richard Nixon's 1968 presidential campaign, the press's access to most candidates has been tightly controlled, with journalists kept at a distance on and off the trail.[43] And as 1988 demonstrated, the less accessible candidate (Bush) was better able to communicate his message than the more accessible one (Dukakis); the kinder and gentler rewards of victory went to the nominee who was better able to keep the pesky media at bay.

However, it is by no means certain that the public now *wants* the press to be in close contact with candidates. In recent years "there has been a significant erosion of public confidence in the press," according to the Times-Mirror Center for the People and the Press.[44] Many journalists, of course, do not need surveys to tell them this; they see it in their mail and telephone calls and they sense it on the road. On the other hand, voters sometimes incorrectly blame the messenger for delivering a disagreeable truth about a well-liked politician. It is the press's special mission to point out that a naked emperor is wearing no clothes, even to a people determined to believe otherwise. Nonetheless, reporters would be foolhardy to ignore emphatic signs of the public's growing displeasure with them.

Consequences for the Presidential Candidates

The two cases of attack journalism examined above provide a reliable indication of a frenzy's consequences for a politician. The rumors of Dukakis's mental impairment certainly took his campaign off its stride and probably played at least some role in his defeat. And the attack on Quayle may have permanently damaged his chance of ever being elected to the presidency. Despite somewhat more positive coverage of Quayle during the 1992 campaign, his press secretary, David Beckwith, sees little likelihood that his boss can overcome the frenzy-generated image burdens anytime soon: "For the indefinite future there will be lingering questions about Quayle based on what people saw or thought they saw in the [1988] campaign, and it's going to be with him for a number of years." [45] Quayle can be certain that remnants of his past frenzy will resurface and develop in his next campaign.

Oddly enough, however, in some ways Quayle's campaign woes helped the man who chose him. While voter concern about the vice-presidential candidate probably cost Bush enough votes in November to deny the GOP ticket a popular-vote landslide, Quayle served as the lightning rod for Democratic and press criticism, deflecting fire from Bush himself. Moreover, Bush's wimp image was transformed into a "take-charge" one overnight as he forcefully defended his running mate and refused to give in to pressure to dump him. In the end, most people vote for president, not vice president. As Bush campaign manager Lee Atwater said, "If the media and the Dukakis campaign had spent nearly as much time trying to strike up a populist theme or trying to develop some real issues instead of going on a rabbit chase after Dan Quayle, they might have drawn blood." [46]

Consequences for Voters

To voters, what seems most galling about attack journalism in presidential election campaigns is not the indignities and unfairness inflicted on candidates, however bothersome they may be. Rather, people often appear to be irate that candidates are eliminated before the electorate speaks, that irreversible political verdicts are rendered by journalists instead of by the rightful jury of citizens at the polls. The press sometimes seems akin to the Queen of Hearts in *Alice's Adventures in Wonderland,* who declares, "Sentences first—verdicts afterwards."

The denial of electoral choice is an obvious consequence of some frenzies, yet the news media's greatest impact on voters is not in the winnowing of candidates but in the encouragement of cynicism. There is no doubt that the media, particularly television, have the power to influence people's attitudes. With the decline of political parties, news publications and broadcasts have become the dominant means by which citizens learn about public officials; and while news slants cannot change most individuals' basic views and orientation, they can dramatically affect *what* people think about and *how* they approach a given subject. [47]

The electorate's media-assisted cynicism has been confirmed in a host of studies and surveys. [48] The Gallup poll rankings of major-party presidential candidates, conducted in the late summer of each election year, are one crude but revealing measure of declining public confidence in political leaders, as reflected in the total number of "very favorable" ratings given to the Democratic and Republican nominees:

1952	84%	1972	63%
1956	94%	1976	69%
1960	77%	1980	51%
1964	76%	1984	68%
1968	63%	1988	42%

By no means can this trend be ascribed solely to media coverage, as disillusioning events (Vietnam, Watergate, recessions, the Iran-Contra affair, and others) crowd the post-1964 calendar. But this period of growing cynicism also coincides almost precisely with the new era of freewheeling journalism. And surely the kind of coverage these defining events received made a difference in the public's interpretation of them.

Consequences for the Political System

The enhanced—some would say inordinate—influence of the contemporary press is pushing the American political system in certain unmistakable directions. On the positive side are the increased openness and accountability visible in government and campaigns during the last two decades. This is balanced by two disturbing consequences of modern press coverage: the trivialization of political discourse and the dissuasion of promising presidential candidacies.

As to the former, the news media have had plenty of company in impoverishing the debate, most notably from politicians and their television consultants. Nonetheless, journalists cannot escape some of the responsibility. First, the press itself has aided and abetted the lowering of the evidentiary standards held necessary to make a charge stick. In addition to the publication of rumor and the insinuation of guilt by means of innuendo, news outlets are willing to target indiscriminately not just real ethical problems, but possible problems and the perception of possible problems. Second, the media often give equal treatment to venial and mortal sins, rushing to make every garden-variety scandal another Watergate. Such behavior not only engenders cynicism, but also cheapens and dulls the collective national sense of moral outrage that ought to be husbanded for the real thing. Third, the press often devotes far more resources to the insignificant gaffe than to issues of profound national and global impact. On many occasions, peccadilloes have supplanted serious debate over policy on the front pages.

The second troubling consequence of media coverage has to do with the recruitment of presidential candidates.[49] Simply put, the price of power has been raised dramatically, too high for some outstanding potential officeholders.[50] An individual contemplating a run for office must now accept the possibility of almost unlimited intrusion into his or her financial and personal life. Every investment made, every affair conducted, every private sin committed from college years on may one day wind up on television or in a headline. For a reasonably sane and moderately sensitive person, this is a daunting realization, with potentially hurtful results not just for the candidate but for his or her immediate family and friends. American society today may well be losing the services of many exceptionally talented individuals who could make outstanding contributions to the commonweal, but who understandably will not subject themselves and their loved ones to abusive, intrusive press coverage.

Some journalists argue that concern about the deterrence of good candidates is overdrawn and outweighed by the advantages of close scrutiny. Ted Gup of *Time* magazine's Washington bureau notes, "If people don't want to undergo profound scrutiny, then perhaps they have something to hide or have onion-thin skin and shouldn't be in public office." [51] Adds the *Washington Post*'s Richard Harwood, "Some people may be discouraged from running, but we've never had to call off an election for a lack of candidates." [52]

These arguments are indisputable in part. Frenzies have indeed eliminated some unworthy potential presidents, and they certainly have measured crisis-managing skills. But a long campaign is fraught with enough revealing tests of a politician's mettle; it is wholly unnecessary to manufacture any on tangential matters. And while no ballot will ever be devoid of names, some may well be lacking good ones if the price of power continues to escalate. Again it comes down to what types of persons our society wishes to nudge toward and away from public service. As *New York Times* columnist Anthony Lewis suggests, "If we tell people there's to be absolutely nothing private left to them, then we will tend to attract only those most brazen, least sensitive personalities. Is that what we want to do?" [53]

Fortunately, we have not yet reached the point where only the brazen enter public service, but surely the emotional costs of running for office are rising. Intensified press scrutiny of private lives and the publication of unsubstantiated rumors have become a major part of this problem. After every election cycle, reflective journalists express regret for recent excesses and promise to do better, but sadly the abuses continue. No sooner had the 1992 presidential campaign begun in earnest than Democratic front-runner Bill Clinton was sidetracked for a time by unproven allegations from an Arkansas woman, Gennifer Flowers, about an extramarital affair. The charges were initially published in a supermarket tabloid, the *Star*, and while some news outlets at first downplayed the story because of the questionable source, others ballyhooed it so extravagantly that Clinton was forced to respond, thus legitimizing full coverage by virtually all news organizations.

This classic case of lowest-common-denominator journalism guaranteed the continued preeminence of the character issue for yet another presidential campaign cycle, and in many ways the situation frustrated reporters and voters alike. Both groups can fairly be faulted for this trivialization of campaign coverage: reporters for printing and airing unproven rumors, and voters for watching and subscribing to the news outlets that were the worst offenders. But journalists and their audiences also have it within their power by means of professional judgment and consumer choice to change old habits and bad practice.[54] Hope springs eternal . . . and in the meantime, attack journalism flourishes.

Notes

1. The press explosion is observed in government as well as in campaigns. President Reagan's press spokesman, Larry Speakes, compares the dozen reporters on hand at the White House for Harry Truman's 1945 announcement of the atomic bombing of Hiroshima with the three thousand reporters on a "South Lawn platform half as long as a football field" for the 1981 return of the American hostages from Iran.
2. Reporters T. R. Reid and Lloyd Grove of the *Washington Post* have defined the ratio of media members to the general public as the "hype index." The higher the index number, the greater the hype.
3. John and Mary R. Markle Foundation, *Report of the Commission on the Media and the Electorate: Key Findings* (New York: Markle Foundation, May 1990), 16-18. See also Thomas E. Patterson, "The Press and Its Missed Assignment," in Michael Nelson, ed., *The Elections of 1988* (Washington, D.C.: CQ Press, 1989), 98; Marjorie R. Hershey, "The Campaign and the Media," in Gerald M. Pomper et al., eds., *The Election of 1988: Reports and Interpretations* (Chatham, N.J.: Chatham House, 1989), 96-98; James Glen Stovall, "Coverage of the 1984 Presidential Campaign," *Journalism Quarterly* 65 (Summer 1988): 444-445; Thomas E. Patterson, *The Mass Media Election* (New York: Praeger, 1980); and Michael J. Robinson, Nancy Conover, and Margaret Sheehan, "The Media at Mid-Year for McLunites," *Public Opinion* 3 (June/July 1980): 41-45.
4. Robert S. Litcher, "How the Press Covered the Primaries," *Public Opinion* 11 (July/August 1988): 45-49.
5. Tom Rosenstiel, *Los Angeles Times*, telephone interview with author, October 30, 1989.
6. Robert Novak, interview with author, Washington, D.C., August 25,

1989.
7. Jim Gannon, *Detroit News,* interview with author, Charlottesville, Va., September 28, 1989.
8. David S. Broder, *Behind the Front Page* (New York: Simon & Schuster, 1987), 238-239.
9. Ted Koppel, interview with author, Washington, D.C., January 5, 1990.
10. Steven Roberts, *U.S. News & World Report,* interview with author, Washington, D.C., August 18, 1989.
11. Novak interview.
12. Mark Shields, interview with author, Washington, D.C., September 22, 1989.
13. See the journalists quoted by John B. Judis, "The Hart Affair," *Columbia Journalism Review* 25 (July/August, 1987): 21-25.
14. Roberts interview.
15. Michael Barone, *U.S. News & World Report,* interview with author, Washington, D.C., September 7, 1989.
16. American Society of Newspaper Editors, *The Changing Face of the Newsroom* (Washington, D.C.: ASNE, May 1989), 33; William Schneider and I. A. Lewis, "Views on the News," *Public Opinion* 8 (August/ September 1985): 6-11, 58-59; and Robert Lichter, Stanley Rothman, and Linda S. Lichter, *The Media Elite* (Bethesda, Md.: Adler and Adler, 1986). Note, however, that the Lichter sample was probably weighted disproportionately toward the most liberal segment of journalism. See Robert M. Entman, *Democracy without Citizens: Media and the Decay of American Politics* (New York: Oxford University Press, 1989), 49-50.
17. See Dom Bonafede, "Crossing Over," *National Journal* 21 (January 14, 1989): 102; Richard Harwood, "Tainted Journalists," *Washington Post,* December 4, 1988, L6; Charles Trueheart, "Trading Places: The Insiders Debate," *Washington Post,* January 4, 1989, D1, 19; and Kirk Victor, "Slanted Views," *National Journal* 20 (June 4, 1988): 1512.
18. "*Roe v. Webster,*" *Media Monitor* 3 (October 1989): 1-6.
19. Richard Harwood, "A Weekend in April," *Washington Post,* May 6, 1990, B6. See also David Shaw, "Abortion and the Media," (four-part series), *Los Angeles Times,* July 1, 1990, A1, 50-51; July 2, 1990, A1, 20; July 3, 1990, A1, 22-23; July 4, 1990, A1, 28-29.
20. "Post Haste," *New Republic* 201 (December 4, 1989): 9-10. *Post* editors forbade employees from engaging in this practice after the first reported occurrence of it, but to no avail.
21. The importance of the agenda-setting function is discussed throughout Shanto Iyengar and Donald R. Kinder, *News That Matters: Television and American Opinion* (Chicago: University of Chicago Press, 1988). See especially the conclusions reached on 4, 33.
22. David Whitman, "Who's Who Among the Homeless," *New Republic* 199 (June 6, 1988): 18-20. See also *Washington Post,* April 19, 1989, D10.
23. David Gergen, "The Message to the Media," *Pubic Opinion* 7

(April/May 1984): 5-6.

24. Patrick J. Buchanan, "Pundit vs. 'Re-pundit' on Writers' Rights and Reasons," *Richmond Times-Dispatch*, December 29, 1988, A14.
25. Michael J. Robinson and Maura Clancey, "General Election Coverage: Part 1," *Public Opinion* 7 (December/January 1985): 49-54, 59.
26. See Robert S. Lichter, Daniel Amundson, and Richard E. Noyes, "Election '88 Media Coverage," *Public Opinion* 11 (January/February 1989): 18-19, 52; and Eleanor Randolph, "CBS Hanging Tough on Vice President," *Washington Post*, February 13, 1988, A15.
27. See Robert S. Litcher, Daniel Amundson, and Richard E. Noyes, *The Video Campaign: Network Coverage of the 1988 Primaries* (Washington, D.C.: American Enterprise Institute, 1988).
28. This is sometimes termed "structural bias."
29. See, for example, William C. Adams, " '84 Convention Coverage," *Public Opinion* 7 (December/January 1985): 43-48.
30. Dennis King, *Lyndon LaRouche and the New American Fascism* (Garden City, N.Y.: Doubleday, 1989). See especially 121-122.
31. Rowland Evans and Robert Novak, "Behind Those Dukakis Rumors," *Washington Post*, August 8, 1988, A13. Reporters from six major news organizations (all three networks, the *Washington Post*, *U.S. News & World Report*, and the *Los Angeles Times*) told us they had been contacted by Bush operatives about the rumor, and they knew of colleagues at other outlets who had also been called. See also Thomas B. Rosenstiel and Paul Houston, "Rumor Mill: The Media Try to Cope," *Los Angeles Times*, August 5, 1988, 1, 18.
32. Roberts interview.
33. See Edward Walsh, "Dukakis Acts to Kill Rumor," *Washington Post*, August 4, 1988, A1, 6.
34. Frank Sesno, interview with author, Charlottesville, Va., September 27, 1989.
35. Gerald M. Boyd, "Doctor Describes Bush as 'Active and Healthy,' " *New York Times*, August 6, 1988.
36. Kirk O'Donnell, telephone interview with author, June 29, 1990.
37. Dennis King, in *Lyndon LaRouche*, 122, commented upon "the usual [media] reluctance to cover anything relating to LaRouche."
38. Nina Totenberg, telephone interview with author, October 4, 1989.
39. The network Quayle coverage on evening news shows, August 18-27, 1988, compiled from Vanderbilt University's *Television News Index and Abstracts* (Nashville, Tenn., August 1988), was as follows:

Network	Quayle stories	Quayle minutes	Lead[a] minutes	Total coverage[b]
ABC	22	49:50	35:00	85.5%
CBS	20	42:50	32:40	67.5
NBC	18	38:20	30:20	68.2

[a] "Lead" means the first item on the evening news.
[b] Coverage of campaigns.

40. Jill Abramson and James B. Stewart, "Quayle Initially Failed a Major Exam at DePauw, Former School Official Says," *Wall Street Journal,* August 23, 1988, 54.

41. Jill Abramson, interview with author, Washington, D.C., August 4, 1989.

42. Jerald terHorst, interview with author, Washington, D.C., August 4, 1990.

43. See Joseph McGinniss, *The Selling of the President 1968* (New York: Trident, 1969).

44. Times-Mirror Center for the People and the Press, "The People and the Press, Part 5: Attitudes toward News Organizations," conducted by the Gallup organization in August to October 1989. For some of the findings of this study, see Larry J. Sabato, *Feeding Frenzy: How Attack Journalism Has Transformed American Politics* (New York: The Free Press, 1991), 202.

45. David Beckwith, telephone interview with author, December 27, 1989. For example, David Broder and Bob Woodward wrote an influential and generally positive series assessing Quayle's career that ran in the *Washington Post* January 5-12, 1992. The series helped to take some of the disparaging edge off Quayle's image.

46. Lee Atwater, interview with author, Front Royal, Va., January 6, 1990.

47. See Iyengar and Kinder, *News That Matters*; Thomas E. Patterson, *The Mass Media Election* (New York: Praeger, 1980); Charles Press and Kenneth VerBurg, *American Politicians and Journalists* (Glenview, Ill.: Scott, Foresman, 1988), 62-66; Shanto Iyengar, Mark D. Peters, and Donald Kinder, "Experimental Demonstrations of the 'Not So Minimal' Consequences of Television News Programs," *American Political Science Review* 76 (December 1982): 848-858; and Roy L. Behr and Shanto Iyengar, "Television News and Real-World Cues and Changes in the Public Agenda," *Public Opinion Quarterly* 49 (Spring 1985): 38-57.

48. See Austin Ranney, *Channels of Power: The Impact of Television on American Politics* (New York: Basic Books, 1983), 75-79.

49. On this general subject, see also Laurence I. Barrett, "Rethinking the Fair Games Rules," *Time* 130 (November 30, 1987): 76, 78; Richard Cohen, "The Vice of Virtue," *Washington Post,* March 10, 1989, A23; Charles Krauthammer, "Political Potshots," *Washington Post,* March 1, 1989; Norman Ornstein, "The *Post*'s Campaign to Wreck Congress," May 29, 1989, A25; and "Ethicsgate," *Wall Street Journal* editorial, July 15, 1983, 26.

50. Increasing intrusiveness and scrutiny are also factors in the lessened attractiveness of nonelective governmental service. See Lloyd M. Cutler, "Balancing the Ethics Code," *Washington Post,* March 13, 1989, A15; Ann Devroy, "Current Climate of Caution: Expanded FBI Checks Slow Confirmations," *Washington Post,* March 13, 1989, A1, 4-5.

51. Ted Gup, interview with author, Washington, D.C., July 10, 1989.

52. Richard Harwood, interview with author, Washington, D.C., July 10,

1989.
53. Anthony Lewis, telephone interview with author, November 20, 1989.
54. Some remedies from the perspectives of both journalists and news consumers are proposed in Sabato, *Feeding Frenzy,* Chapter 8.

5. CAMPAIGNS THAT MATTER

Samuel L. Popkin

Sam Popkin challenges the excoriation by the other essayists in this volume of television's impact on political campaigns. Popkin argues that voters do not need to know everything about candidates and their policy stances to make informed voting decisions. Instead, everyday existence gives voters the means to put into context and to process whatever information television and campaigns throw at them. What matters is not the specific information conveyed by a particular advertisement, endorsement, or image, but how that information conforms to a network of prior observations and experiences the voter already has internalized.

Campaigns, Popkin argues, rely more on symbols than cut-and-dried data to tell voters about candidates. Political advertisements, media consultant Doug Bailey noted, stress "symbols, symbols, symbols. I don't mean just sort of waving the flag. That's the most obvious kind of symbol. And I don't believe, by the way, that there's anything wrong with doing that. But Michael Dukakis in the tank, in the helmet, with the strap, is a symbol. It is ... a symbol of the mismatch of that man with the presidency, a visual symbol, and it also gets to the question of character.... [T]he introduction of television into politics made real the capacity for every voter to make a judgment of character in the broadest sense of that term of the candidate running for the highest office in the land."

Popkin contends that what matters most is how voters interpret the images and signals they consume from television (and from other sources related to the campaign). The problem for present-day presidential campaigns is not determining the quantity or type of information to provide through advertising and news coverage of campaign events but figuring out whether voters are getting the intended messages from TV ads and news reporting that have become more image-oriented. Sophisticated voters are needed to make accurate inferences about the consequences of electing one candidate over another.

Popkin believes that voters are more than up to the job of using campaign imagery to make meaningful choices. As political consultant Greg Schneiders said about the 1984 Reagan campaign, "there is a lot of content in reminding people . . . about the Morning in America ads, and it's reminding people they feel better about their country today than they did four years ago. That's enormous content." For Popkin, campaigns, not voters, have fallen short. What now is needed, Popkin contends, are bigger, louder, longer presidential campaigns that can provide the volume of information that voters are capable of processing. —Ed.

There is something rather miraculous about the fact that citizens believe that leaders selected by balloting are entitled to govern, even though election campaigns are commonly criticized as tawdry and pointless affairs, full of dirty politics, dirty tricks, and mudslinging. Critics want to turn down the noise level of campaigns and focus them on more important issues. On the other hand, they want more people to vote. However, these goals are contradictory: refined, decorous campaigns will not raise more important issues, mobilize more voters, or overcome offstage mutterings about race and other social issues. Not surprisingly, most proposals for reforming campaigns fail to indicate how the reforms would affect voters or improve the system.

Most reformers do not look closely at how voters respond to campaigns nor at the people they wish to sanitize. Contrary to the reformers' views, voters are not passive victims of mass media manipulators or campaign consultants. They are able to navigate campaigns and make sense of the bits and pieces of information gleaned from these sources. Voters learn from campaigns because they know how to read the media and the politicians—that is, because they reason about what they see and hear. The better we understand voters and how they reason, the more sense campaigns make and the more we see how campaigns matter in a democracy. Indeed, I argue that because campaigns do actually help voters make their ballot choices, we need bigger, necessarily more expensive campaigns to get voters interested and involved in elections.

First I will discuss how voters reason, on the basis of information obtained from daily life as well as from information shortcuts. I will then show how campaigns help voters to make choices and why campaigns face a bigger job today. Finally, I will demonstrate how bigger campaigns can help voters and why the arguments that campaigns are either irrelevant or misleading are wrong.

How Voters Reason

It is easy to demonstrate that Americans have limited knowledge of basic textbook facts about their government and the political debates of the day.[1] However, evaluating citizens only in terms of factual knowledge is a misleading way to assess their competence as voters because it fails to acknowledge the importance of using information shortcuts. These easily obtained and used forms of information serve as second-best substitutes for harder to obtain kinds of data, providing rules of thumb that help voters to evaluate information and to simplify the process of choosing a candidate.

In general, voters do not devote much time or energy to deciding how to vote. This is easy to understand. In economic terms, the process of procuring, analyzing, and evaluating information carries a cost—the investment of time and energy. And the expected return is small relative to that obtained from other uses of the same time, far smaller than for decisions about personal consumption. The resources expended to gather and process information before making the latter decisions have a direct effect on the quality of the outcome for the consumer, whereas time and money spent gathering information about candidates may lead to a better vote, but not necessarily to a better outcome. For example, time spent deciding where to travel may lead to better vacations, whereas time spent evaluating economic policies may not lead to better policies but only to a better-informed vote. Some people, of course, find politics so fascinating that they gather information even when they have no personal stake in political outcomes, but in general everybody's business is nobody's business.

Voters, however, have far more data about politics than measurements of their textbook knowledge would suggest. Although they may have few incentives to gather information in order to become good citizens, and thus may be uninformed about many issues, they do apply to political decision making a great deal of information acquired as a *byproduct* of their ordinary activities. For example, a businessperson interested in selling computers overseas learns about technology transfer laws to develop sales plans. A senior citizen learns about Social Security benefits to make his or her retirement plans. Home builders and prospective home buyers learn about interest rates to estimate the cost of properties.

Despite the many kinds of information that voters acquire in daily life, their knowledge of government and politics has large gaps. To overcome these limitations, they use shortcuts. The most familiar, of course, is party identification, which serves as a shortcut to storing and encoding their past experiences with political parties.[2] This includes information about social groups prominent in the party, its priorities, and the performance of the party and its past presidents in various policy areas. This generalized information provides "default values" that enable voters to assess candidates about whom they have no other information. By keeping generalized tallies on issues, they avoid the need to know the specifics of every policy proposal.

Curiously, throughout this century, an antiparty strain of reformism in America has argued for nonpartisan elections. Party labels are viewed as giving voters the illusion of informed choice while allowing them to ignore important differences between candidates on the newly emerging issues of the day. Take away party labels, the reformers argue, and voters will pay attention to the "real" differences between candidates on the issues. In reality, however, voters evaluate candidates and form images of them on the basis of the same types of information shortcuts they use to form views of parties, and issue positions are by no means their only criteria in the evaluation.[3]

Voters use information shortcuts in assessing the personal characteristics of the candidate and his or her issue positions. A candidate's past positions will be used to estimate those in the future. When voters are uncertain about past positions, they will accept as a proxy information about the candidate's personal demographic characteristics and the groups with which he or she has associated. And since voters find it difficult to gather information about the performance of politicians outside their district or state, they will accept campaign competence as a proxy for competence in elected office as an indication of the political skills needed to handle the issues confronting the government.

Thus voters are concerned about personal character and integrity because they generally cannot infer the candidate's true commitments from his or her past votes, most of which are based upon a hard-to-decipher mixture of compromises between ideal positions and practical realities. Evaluations of personal character serve as a substitute for information about past demonstrations of political character.

Voters care about the personal competence of the candidate, his or her ability to deliver benefits. They assess overall competence because they do not understand all the problems a president must deal with and do not make individual judgments about these problems. When assessing competence, they also economize by judging campaign behavior instead of researching the candidate's past governmental performance.

Voters also rely on the opinions of others whom they trust and with whom they discuss political news. These opinions can alert them to matters deserving more than minimal attention.[4] Just as alarms alert firefighters, saving them the effort of patrolling to look for smoke, so do information shortcuts save voters the effort of constantly searching for relevant facts. It makes sense for them to use these shortcuts provided by persons whose evaluations they trust. The development and expansion of the mass media provide voters with the additional shortcut of validating their opinions by reference to national political leaders and commentators whose positions and reputations become known to people over time.[5]

Finally, people use shortcuts when choosing a candidate. Faced with an array of candidates, some much better known than others, and all known in different ways that make comparison difficult, voters will seek a clear and accessible criterion for comparing them. This usually means looking for the sharpest differences between the candidates, which are then incorporated into narratives that make it possible to compare the candidates without spending the time and energy needed to make independent evaluations of them. When people compare candidates on the basis of the most obvious differences rather than the most important, they are conducting the equivalent of a Drunkard's Search, looking for their lost car keys under the streetlight only because it is easier to search there.[6]

As a shortcut in assessing a candidate's future performance without collecting more data, people assemble what information they have into a causal narrative, that is, a story. Because a story needs a main character, they can create one from their knowledge of persons with traits or characteristics like those of the candidate. This allows them to go beyond their incomplete information and to hold together and remember more data than they otherwise could. Because these stories are causal narratives, they allow voters to think about government in causal terms and to evaluate what it will do. Narratives thus help people incorporate their reasoning about government into projections about candidates; their assumptions "confer political significance on some facts and withhold it from others."[7]

Thus low-information reasoning is by no means devoid of substantive content; it is a process that economically incorporates learning and information from past experiences, daily life, the media, and political campaigns. As Tony LaRussa, the manager of the Oakland Athletics, put it: "When you trust your gut you are trusting a lot of stuff that is there from the past."[8]

The use of information shortcuts is an inescapable fact of life and will occur no matter how educated we are, how much information we have, or how much thinking we do. The belief that increased education would lead to higher levels of textbook knowledge about government, and that this knowledge would in turn enable the electorate to measure up to its role in democratic theory, was misplaced. Education doesn't change *how* we think. Educated voters still sample the news, and they still rely on shortcuts and calculation aids in assessing information, assembling scenarios, and making their choices. Therefore, improving democracy depends upon improving the shortcuts people use in considering campaigns, not upon getting citizens to stop using them.

Taking the high road and trying to eliminate shortcuts will work no better than promoting more civic education.

Why Campaigns Matter

Campaigns give voters much of the information used in their reasoning as they deal with uncertainty about the immediate effect on them of governmental actions. The types of issues commonly dealt with in campaigns—bounced checks, congressional salary increases, welfare cheats, and $600 ashtrays—become important symbols in politics because they are so easily grasped that candidates can use them as focal points to organize debate and keep memories of a position alive long after the specifics have been forgotten. A $600 ashtray purchased by the Pentagon will receive more attention from politicians and the media during campaigns than programs costing billions of dollars. Voters, as well as politicians, find it easier to judge the value of ashtrays and pay raises than to decide whether a multibillion dollar electronics project, such as a cruise missile or a weather satellite, is fairly priced, well managed, or necessary.

Campaigns attempt to achieve a common focus, to make one distinction paramount in voters' minds. They try to develop a message for a general audience, a call that will reach beyond the "disinterested interest" of the educated and highly attentive, on the one hand, and the narrow interests of "issue publics"—subsets of the overall public that care a great deal about a particular issue and therefore are likely to pay attention to it—on the other.[9] Each campaign attempts to organize the many splits within the electorate by setting the political agenda in the way most favorable to its own candidate. As Bernard Berelson and his colleagues observed almost forty years ago:

> At the start of any political campaign the individual voter agrees with some issues, disagrees with others and is indifferent to some. As a result there are potential or actual conflicts over issues within individuals, within social groups and strata and within parties. Then the campaign goes on, and somehow the combination of internal predispositions and external influences brought to bear on the content of the campaign (the issues) leads to a decision on election day that one or the other party shall control the presidency for the following four years.
>
> Thus, what starts as a relatively unstructured mass of diverse opinions with countless cleavages within the electorate is transformed into, or at least represented by, a single basic cleavage between the two sets of partisans . . . disagreements are reduced, simplified and generalized into one big residual difference of opinion.[10]

The transformation of "What have you done for me lately?" into "What have you done for us lately?" is the essence of campaigning. Transforming unstructured and diverse interests into a single coalition, making a single distinction dominant, requires the creation of new constituencies and political identities. It requires the aggregation of countless "I's" into a few "we's."

In the end, candidates manage to get a large proportion of the citizenry sorted into opposing camps, each convinced that the positions and interests of the other side add up to a less desirable package of policies. Thus campaigns, to the extent that they are successful, temporarily change the basis of political involvement from citizenship to partisanship, and in the process attract interest and votes from persons who generally find politics uninteresting or remote.

The Growing Importance of Campaigns

Changes in government, society, the mass media, and the nomination process have made campaigns more important today than they were fifty years ago when modern studies of them began. The electorate consists of many more diverse and overlapping constituencies; a more educated citizenry cares about more issues; there are many more communications channels through which specific segments of the electorate are organized and mobilized; and the primary nominating process leaves parties more divided at the start of the campaign. Each of these changes means that the process of making one distinction paramount in voters' minds is more difficult.

Today there are more government programs—Medicare, Social Security, welfare, and farm supports are obvious examples—and, consequently, more issue publics. Coalitions have organized around policies toward specific countries, such as Israel or Cuba; various conservation and environmental groups; and groups concerned with social issues, such as abortion and gun control.

A wrong prediction in the classic study of the 1948 election, *Voting,* emphasizes how the growth of government, the development of specialized communications channels, and the changes in society since the 1940s interact to create new issue publics. Written in 1954, that book stated, "It would be difficult in contemporary America to maintain strong voting differences by sex, because there are few policy issues persisting over a period of time that affect men and women differently." [11] In the last twenty years, a distinctive women's vote has in fact developed; more women than men have cast votes for Democrats in the last four presidential elections, and there are several issues on which women and men have different attitudes and priorities. They differ, for example, in their evaluations of the effectiveness of the Great

Society programs in the 1960s. The gender gap, as it is called, developed as a result of the growth of single-parent families and an increased government role in social welfare and childcare. *Voting* also concluded that there could be no policy movements based on the special interests of old age, such as pensions, because such a movement would have no continuity over time.[12] Since 1948, of course, we have seen a major migration of retired persons to senior citizens' communities throughout the Sunbelt, as well as the growth of organizations like the American Association of Retired Persons that communicate and transmit information about Social Security and Medicare. And not surprisingly the elderly have higher levels of knowledge than other citizens about specific legislative actions relating to Social Security and Medicare.[13]

Just as the expansion of government programs increases the awareness of many voters about government actions, the increased education level of the populace also affects campaigns, both broadening and segmenting the electorate. Educated voters pay attention to more problems and are more sensitive to connections between their lives and national and international events. The result is a more diffuse electorate, segmented by the very abundance of its concerns into many more electronic communities, groups of people whose important identifications are maintained through media rather than direct, personal contact. Such an electorate is harder to form into coalitions. The more divided an electorate, and the more money available to advocates of specific issues or causes, the more time and communications it takes to assemble people around a single cleavage.

The proliferation of media and channels has also increased segmentation of the electorate. As a result, the general campaign has a bigger job, requiring candidates to transmit more messages through more channels than in the past. When the national political conventions were first telecast in 1952, all three networks showed the exact same picture at the same time because there was only one national microwave relay; today, with the proliferation of cable systems and satellite relays, television can show over 100 channels. A voter can receive feedback about a presidential speech or campaign from a variety of religious, ethnic, liberal, conservative, and lifestyle-oriented channels; by 1992 even a comedy channel was offering commentary on state of the union messages. Fifty years ago the only news services were the Associated Press and United Press International; in the 1960s and 1970s papers like the *New York Times* and the *Washington Post* established news services with their own emphases. As channels and options have proliferated, and as commuting time has increased and two-career families have become more common, the proportion of people watching mainstream networks and their news programs has also dropped. This makes the job of integrating segments of the electorate more difficult and expensive.

There are also a great many more specialized radio and TV programs and channels, magazines, newsletters, and even computer bulletin boards that enable people to keep in touch with others holding similar views outside their immediate neighborhoods or communities. Such extended groups are not new, and modern communication technology is not necessary to mobilize them, as a look back at the abolitionist and temperance movements reminds us; however, more channels to mobilize such groups are available today, and the groups they have nurtured are more numerous.[14]

With the development of the presidential primary system, many voters now begin the presidential campaign after having opposed their party's nominee during months of public battle for the nomination; many other members of the party will have spent months listening to intraparty opponents raise questions about the nominee they supported. This complicates the job confronting campaigns, which must deal with the additional task of closing ranks after the factions within the party have been pitted against each other. Today most people learn about their party's nominee in the course of year-long intraparty battles; fifty years ago most voters first learned about the nominees when they received their party's nomination.

Primaries and the expansion of communication media have increased voters' exposure to individual candidates, particularly to personal information about them. This increases the importance of campaigns because it gives voters more opportunities to abandon views based on party default values in favor of views based on candidate information. There are also more opportunities to shift from views based on a candidate's past record to views based on his or her campaign image.

Finally, the more that primaries divide parties, the more cues are needed to reunite parties and remind supporters of losing candidates about their differences with the other party. This means that conventions are now an important part of the campaign rather than merely a prelude to it. Conventions no longer deliberate and choose candidates; instead, they present the electorate with important cues about the social composition of the candidate's coalition and about the candidate's political history and relations with the rest of the party.

More Campaigning

Successful reform is possible only if we begin by examining how voters actually reason and then alter the system to provide more of the information and stimulation they need to reduce the chance of unacceptable outcomes. If voters engage in a Drunkard's Search for the keys to political choice, then campaigns should give them more, not fewer, streetlights to look under. The place where the candidates are shining their lights may not be the best spot to find information about the issue of greater concern to a voter. But at least when people look where the candidates are shining their lights, they are likely to find some electoral key.

I believe that voter turnout has declined between 1960 and 1990 because campaign stimulation, from the media and personal interaction, is also low and declining, and there is less interaction between the media and the grass-roots, person-to-person aspects of voter mobilization. The lack of campaign stimulation is also responsible for the large gap in this country between the turnout of educated and uneducated voters.

Political parties used to spend a large portion of their resources bringing people to rallies. By promoting the use of political ideas to bridge the gap between the individual "I" and the party "we," they encouraged people to believe that they were "a link in a chain" and that the election outcome would depend on what people like themselves chose to do.[15] Today less money and fewer resources are available for rallies as a part of national campaigns. Parties cannot compensate for this loss with more door-to-door canvassing, and in the neighborhoods where it would be safe to walk door to door, no one would be home!

Some of the social stimulation that campaigns used to provide in rallies and door-to-door encounters can be provided by extensive canvassing. This is still done in Iowa and New Hampshire, the first caucus and primary states, where candidates have the time and resources to do extensive personal campaigning and have organizations that telephone people and discuss the election. People contacted by one candidate pay more attention to all the candidates and to the campaign events reported on television and in the papers, which makes them more aware of differences between the candidates. As this awareness grows, they become more inclined to vote.[16]

This suggests a surprising conclusion: the single best way to increase turnout, and thereby increase the capacity of government to tackle our problems, might be to increase spending on campaign activities that stimulate voter involvement. There are daily complaints about the cost of American elections, and certainly the corrosive effects of corporate fund raising cannot be denied; however, it is not true that American elections are costly in comparison with those in other countries.[17] Comparisons are difficult, especially since most countries have parliamentary systems, but it is worth noting that reelection campaigns for the Japanese Diet, equivalent to our House of Representatives, cost over $1.5 million per seat.[18] That would be equivalent to $3.5 million per congressional reelection campaign instead of the current U.S. average of about $400,000 (given the fact that Japan has one-half the U.S. population and 512 legislators instead of 435). Indeed scholars estimate that Diet elections cost between $50 and $100 per constituent, while incumbent members of Congress spend an average of one dollar per constituent. Although the differences in election systems and rules limit the value of such comparisons, it is food for thought that a country with a self-image so different from that of the United States spends so much more on campaigning.[19]

More campaign competition among the candidates, and more voter exposure to this competition, would allow them more time to move from personal character to general political competence and to introduce questions about the past record of the candidates. In the 1984 Democratic primary, Walter Mondale had to ask, "Where's the Beef?" to force Gary Hart to come to grips with specific policy proposals. The more time and stimulation in such campaigns the better.

If government is going to be able to solve our problems, we need bigger and noisier campaigns to rouse voters. It takes bigger, costlier campaigns to sell health insurance than to sell the death penalty; it takes more time and money to communicate about complicated issues of governance than to communicate about racism. The cheaper the campaign, the cheaper the issue.

Campaigns cannot deal with any substance if they cannot get the electorate's attention and interest people in listening to their music. Campaigns need to make noise. Restricting television news to the MacNeil-Lehrer format and requiring all the candidates to model their speeches on the Lincoln-Douglas debates won't solve America's problems.

Objections and Answers

There are two important criticisms of a policy to encourage bigger campaigns. The first is the "spinmaster" objection: contemporary political campaigns are beyond redemption because strategists have become so adept at manipulation that voters can no longer learn what the candidates really stand for or intend to do. Significantly, this conclusion is supported by two opposing arguments about voter behavior. One is that voters are staying home because they have been turned off by fatuous claims and irrelevant advertising. The other is that people are being manipulated with great success by unscrupulous campaign advertising, so that their votes reflect more concern about Willie Horton, school prayer, or flag burning (the so-called issues of the 1988 election) than about widespread poverty, the banking crisis, or global warming.

The second objection is that popular concern about candidates and government in general has been trivialized, so that candidates fiddle while America burns. In the various versions of this critique, voter turnout has descended from its postwar highs because today's political contests are waged over small differences on trivial issues. While Eastern Europe plans a future of freedom under eloquent leaders like Vaclav Havel, and while Mikhail Gorbachev declares an end to the cold war, releases Eastern Europe from Soviet control, and tries to free his countrymen from the yoke of doctrinaire communism, in America Tweedledee and Tweedledum argue about who loves the flag more while Japanese businesses buy Rockefeller Center, banks collapse, the deficit grows, and the holes in the ozone layer get bigger.

Both of these critiques argue that contemporary campaigns are trivial and irrelevant, that advertising and even the candidates' speeches are nothing but self-serving puffery and distortion. This general argument is especially appealing to better-educated voters and the power elite; campaign commercials remind no one of the Lincoln-Douglas debates, and today's bumper stickers and posters have none of the resonance of the Goddess of Democracy in Tiananmen Square or Boris Yeltsin astride a tank. But elitist concerns are not the test of this argument; the test is what voters learn from campaigns. There is ample evidence that voters *do* learn from campaigns. Of course, each party tries hard to make itself look better than the other side. Nonetheless, voter perceptions about the candidates and their positions are more accurate at the end of campaigns rather than at the beginning, and exposure to campaigns is still the best assurance that their perceptions are accurate.

Campaign focus on an issue leads to less voter misperception, not more. When the stakes are raised and more information becomes available, voters acquire more accurate perceptions. The more they care about an issue, the better they are able to understand it; and the more strongly the parties differ on an issue, and the more voters hear about it, the more accurate their perceptions become. There is no denying that misperception is always present in campaigns, but it is also clear that communications do affect choices, making voters more, not less, accurate in their perceptions about candidates and issues.

Despite cries that campaigns have become less substantial in the television era, recent research has supported the findings given in *Voting* that campaign communications increase the accuracy of voters' perceptions. Pamela Johnston Conover and Stanley Feldman, examining the 1976 general election, found that projection, or false consensus, declined as learning proceeded during the campaign; misperceptions occurred "primarily when there [was an] absence both of information and *strong* feelings about the candidate." [20] Contradicting the claim that voters' perceptions are "largely distorted by motivational forces," Jon Krosnick found that voters in the 1984 election were "remarkably accurate in their perceptions of where presidential candidates [stood] relative to one another on controversial policy issues." [21] And in 1988 it was not misperception about Michael Dukakis but the popularity of Ronald Reagan that was critical for George Bush; the lesson of 1988, Alan Abramowitz has concluded, is that "what matters most [is] . . . what issues the voters are thinking about." [22]

In addition to reducing misperception, campaign information also helps people connect issues to government and parties. In 1984, according to Abramowitz, Lanoue, and Ramesh, "Voters who followed the presidential campaign closely were more likely to connect their personal financial situation with macroeconomic trends or government policies. . . . Attributions of responsibility for changes in economic well-being are based in part on cues received from the political environment and particularly from the mass media." [23] Thus in a world in which causal reasoning matters but voters have only limited knowledge of government, campaign communications heighten awareness of how government affects their lives while reinforcing policy differences between parties and candidates.

The mind of the voters is not a tabula rasa to be easily programmed and manipulated. Voters remember past campaigns and presidents, past performance that failed to match promise. They have a sense of who is with them and who is against them; they make judgments about unfavorable news, editorials, and advertisements from hostile sources, ignoring some of what is favorable to those they oppose and some of what is negative to those they support. In managing their personal affairs and making decisions about their work, they collect information that can be used as a reality test against campaign claims and media stories. They notice the difference between behavior that has real consequences and mere talk.

Big brother is not gaining on the public. This is a finding to keep in mind at all times, for many criticisms of campaigns simplistically assume that voters are misled because politicians and campaign strategists have manipulative intentions. This assumption is not borne out by the evidence; voters know how to read the media and the politicians better than most media critics acknowledge.

Critics of campaign spinmasters (and of television in general) are fond of noting that campaigners and politicians intend to manipulate and deceive, but they wrongly credit them with more success than they deserve. As Michael Schudson has noted, in the television age, whenever a president's popularity has been high, it has been attributed to unusual talents for using television to sell his image. He notes, for example, that in 1977 the television critic of the *New York Times* called President Carter "a master of controlled images," and that during the 1976 primaries David Halberstam wrote that Carter "more than any other candidate this year has sensed and adapted to modern communications and national mood. . . . Watching him again and again on television I was impressed by his sense of pacing, his sense of control, very low key, soft." [24] A few years later this master of images still had the same soft, low-key voice, but now it was interpreted as indicating not quiet strength but weakness and indecision.

As these examples suggest, media critics are generally guilty of using one of the laziest and easiest information shortcuts of all. Assuming that a popular politician is a good manipulator of the media or that a candidate won because of his or her media style is no different from what voters do when they evaluate presidents by reasoning backward from known results. The media need reform, but so do the media critics. One cannot infer, without astonishing hubris, that the American people have been successfully deceived simply because a politician wanted them to believe his or her version of events. But the media critics who analyze political texts without any reference to the actual impact of the messages do just that. [25]

Negativism and Trivia

The second objection is that campaigns are sideshows in which voters amuse themselves by learning about the differences between irrelevant candidates who debate minor issues while the country stagnates and inner cities explode. Many also assume that the negativism and pettiness of candidates' attacks on each other encourage a "pox on all your houses" attitude. These attitudes suggest a plausible hypothesis, which can be tested in a simple experiment, one that can be thought of as a "Stop and Think" experiment because it is a test of what happens if people stop and think about what they know of the candidates and issues and tell someone what they know. [26] First, ask a random sample of people across the country what they consider to be the most important issues facing the country, then ask them where the various candidates stand on these issues. Next, ask them to state their likes and dislikes about the candidates' personal qualities and issue stands and about the state of the country. After the election, find out whether these interviewees were more or less likely to vote than people who were not asked to discuss the campaign. If the people interviewed voted less often than those not interviewed, there is clear support for the charge that trivia, negativism, and irrelevancy are turning off the American people and suppressing turnout.

In fact, the National Election Studies done by the University of Michigan's Survey Research Center, which houses the Center for Political Studies, constitute precisely such an experiment. In every election since 1952, people have been asked what they care about, what the candidates care about, and what they know about the campaign. After the election, people have been interviewed again and asked whether they voted; then the actual voting records have been checked to see whether the respondents did indeed vote. The results convincingly demolish the trivia and negativism hypothesis. In every election, people who have been interviewed were more likely to vote than other Americans.[27] Indeed, the expensive and difficult procedure of verifying turnout against voting records by personal reinterview was initiated because scholars were suspicious of the fact that the turnout reported by respondents was so much higher than the actual turnout of all Americans as well as the turnout reported in surveys conducted after the election. Thus respondents in the National Election Studies, after two hours of thinking about the candidates, the issues, and the campaign, were more likely to vote than other people and also more likely to try to hide that they did not vote! Moreover, if people are reinterviewed in later elections, their turnout continues to rise. And, while an interview reduces nonvoting in a presidential election by up to 20 percent, an interview in a local primary may reduce nonvoting by as much as half.[28]

When people put the pieces together they are turned on, not turned off. Indeed the 1991 Louisiana gubernatorial election between Edwin Edwards and David Duke demonstrate this; people will turn out in record numbers to choose between an unabashed rogue and a former Klansman as well as Nazi sympathizer when the campaign is big enough to keep people mobilized.

As Peggy Noonan has noted of the 1988 campaign, "There should have been more name-calling, mud-slinging and fun. It should have been rock-'em-sock-'em the way great campaigns have been in the past. It was tedious." [29] Campaigns cannot deal with any substance if they cannot get the electorate's attention and interest people in listening to their music. Campaigns need to make noise. The use of sanitary metaphors to condemn politicians and their modes of communication say more about the distaste for American society of the people using these metaphors than about the failings of politicians.

Campaigns are essential in any society, particularly one that is culturally, economically, and socially diverse. If voters look for information about candidates under streetlights, that is where candidates must campaign, and the only way to improve elections is to add streetlights. Reforms only make sense if they are consistent with the gut rationality of voters. Ask not for more sobriety and piety from citizens for they are voters not judges; offer them instead cues and signals that connect their world with the world of politics.

Conclusion

Campaigns are blunt instruments, not scalpels. They are for ratifying broad changes in direction that have been worked out between campaigns or for rejecting incumbents and their policies. The reformist hope that campaigns can raise new and complicated issues or bring Americans to a deeper understanding of the most complex issues facing the country is misguided. Campaigns must mobilize the less interested, and they cannot do that while simultaneously making the discourse complex enough to satisfy the most attentive reformers.

Before we attempt to take the passions and stimulation out of politics, we ought to be sure that we are not removing the lifeblood as well. The challenge to the future of American campaigns, and hence to American democracy, is how to bring back the brass bands and excitement in an age of electronic campaigning. Today's campaigns have more to do because an educated, media-centered society is a broadened, segmented electorate that is harder to rally, while today's campaigns have less money and fewer troops with which to fight their battles.

Notes

1. Michael X. Delli Carpini and Scott Keeter, "Political Knowledge of the U.S. Public: Results from a National Survey," paper given at the annual meeting of the American Association for Public Opinion Research, May 1989.
2. Morris Fiorina, *Retrospective Voting in American National Elections* (New Haven: Yale University Press, 1979).
3. Charles Adrian, "Some General Characteristics of Nonpartisan Elections," *American Political Science Review* 46 (September 1952): 766-776.
4. Mathew McCubbins and Thomas Schwartz, "Congressional Oversight Overlooked: Police Patrols versus Fire Alarms," *American Journal of Political Science* 2 (February 1984): 165-179.
5. See also the discussion in Alan Houston, "Walter Lippman: Public Opinion in the Great Society," unpublished manuscript, 1984.
6. Abraham Kaplan, *The Conduct of Inquiry* (San Francisco: Chandler Publishing Company, 1964), 11, 17-18.
7. Donald R. Kinder and Walter R. Mebane, Jr., "Politics and Economics

in Everyday Life," in Kristen R. Monroe, ed., *The Political Process and Economic Change* (New York: Agathon Press, 1982), 146.

8. Quoted in George Will, *Men at Work* (New York: Macmillan, 1990), 30.

9. Bernard Berelson, Paul Lazarsfeld, and William McPhee, *Voting: A Study of Opinion Formation in a Presidential Campaign* (Chicago: University of Chicago Press, 1954), 27-28, 30; the phrase "disinterested interest" is theirs.

10. Ibid., 183.

11. Berelson, Lazarsfeld, and McPhee, *Voting*, 74.

12. Ibid.

13. John Zaller, "Measuring Individual Differences in Likelihood of News Reception," unpublished paper.

14. Over the past fifty years, as surveys have become increasingly available to study public opinion, knowledge about voting and elections has increased. There have also been losses, as national surveys have replaced the detailed community orientation of the original Columbia University studies. We know much more about individuals and much less about extended networks, and we have not adequately examined the implications for society and campaigning of the transition from face-to-face to electronic communities.

15. George A. Quattrone and Amos Tversky, "Causal Versus Diagnostic Contingencies: On Self-Deception and the Voter's Illusion," *Journal of Personality and Social Psychology* 46 (1984), 237-248, develop the concept of the voter's illusion. Voters' illusions and the "link in a chain" arguments of political organizers and collective action theorists are based upon the same logic.

16. Samuel L. Popkin, "The Iowa and New Hampshire Primaries: The Interaction of Wholesale and Retail Campaigning," paper presented at the annual meeting of the American Political Science Association, San Francisco, Sept. 1, 1990.

17. Howard R. Penniman, "U.S. Elections: Really a Bargain?" in *The Mass Media in Campaign '84: Articles from Public Opinion Magazine*, ed. Michael J. Robinson and Austin Ranney (Washington, D.C.: American Enterprise Institute, 1985).

18. Charles Smith, "Captives of Cash," *Far Eastern Economic Review*, March 9, 1989, 16-21; David Sanger, "Will Japan Scandal Force Changes in the System?", *New York Times*, April 27, 1989. When told how much more is spent in Japan, many people appear to change their opinions instantly about the lavishness of campaign spending in the U.S.; this says much about perceptions of America's status.

19. Note further that the Japanese money is all spent on contributions, funeral wreaths, organizations, and the like because there is no television advertising by candidates.

20. Pamela Johnston Conover and Stanley Feldman, "Candidate Perception in an Ambiguous World: Campaigns, Cues, and Inference Processes," *American Journal of Political Science* 33 (November 1989): 935.

21. Jon A. Krosnick, "Americans' Perceptions of Presidential Candidates: A Test of the Projection Hypothesis," *Journal of Social Issues* 46 (Summer 1990): 159-182.
22. Alan I. Abramowitz and Jeffrey A. Segal, "Beyond Willie Horton and the Pledge of Allegiance: National Issues in the 1988 Elections," *Legislative Studies Quarterly* 15 (November 1990): 565-580.
23. Alan I. Abramowitz, David J. Lanoue, and Subha Ramesh, "Economic Conditions, Causal Attributions, and Political Evaluations in the 1984 Presidential Election," *Journal of Politics* 50 (November 1988): 860.
24. Michael Schudson, "Trout and Hamburger: Telemythology and Politics," *Tikkun* 6 (March-April 1991): 47-51, 86-87.
25. The inside cover for Joe McGinniss's *The Selling of the President,* chronicling Richard Nixon's 1968 campaign, included Alastair Cooke's praise of McGinniss: "His frankness would be brutal if his perceptions of the inherent fraud were not so acute," and Murray Kempton's comment that Mr. Nixon's staff talked to McGinniss "with the candor possible only to persons who have no idea how disgusting they are." Joe McGinniss, *The Selling of the President* (New York: Pocket Books, 1974).
26. The term "Stop and Think" effects is from John Zaller, "Political Competition and Public Opinion," unpublished manuscript, University of California, Los Angeles, Department of Political Science, 1990. Ultimately, I believe, Zaller's work on the effects of one-sided and two-sided communications will make it possible to ascertain just how much of the decline in voter turnout reflects a decline in media stimulation and how much a decline in interpersonal use of political information.
27. Several of the initial studies of the turnout rates of NES respondents are reported in Donald R. Kinder and David O. Sears, "Public Opinion and Political Action," in *The Handbook of Social Psychology,* ed. Gardner Lindzey and Elliot Aronson (Hillsdale, N.J.: Random House, 1985), 703. Verified turnout among survey respondents has been higher than for the rest of the country in every election since verification began in 1964.
28. Ibid., 703.
29. Ibid., E5.

CONCLUSION

Mathew D. McCubbins

Since man first climbed down out of the trees, pundits have been bemoaning changes in society and technology. No doubt some made a living on the rubber mastodon circuit predicting that the fall of Stone Age civilization would result from the discovery of fire. In America the invention of the steam locomotive and the automobile were thought to presage the collapse of American civilization. In the realm of communications, the development of the dime novel in the nineteenth century and of the comic book, swing music (and later rock 'n' roll, not to mention heavy metal and punk rock), and radio in this century were all viewed as spelling doom for our way of life. It was thought that these changes in technology would spawn ignorance and illiteracy.

Whenever technological innovations have offered manufacturers new labor-saving production processes, the workers whose jobs are being eliminated naturally see a threat to their own welfare. It is perhaps only natural that prophets of doom should arise to interpret such changes as sinful incursions against the natural order of things; such prophets find receptive audiences when the cost of transition to a new order is concentrated on only one factor of production. It is not without reason that Joseph Schumpeter characterized capitalism as a system of "creative destruction." [1]

Similarly, when workers found ways to coalesce into labor unions, employers saw red. Whenever peasants rose up to restructure property rights with feudal lords, the nobility saw the handiwork of the devil. Indeed, in almost every case of innovation in economic production, from machinery to principles of organization, where transition to a new equilibrium takes real time and therefore means that some parts of society suffer (at least) short-term welfare losses, the losers in those redistributive processes have seen fit to appeal to moral theories.

Communications Technology and Morality

Luddism has not been confined to the realms of production technologies and property rights. Indeed, moral arguments may be the loudest against technological changes in the way information is exchanged. Each advance, from pictograms to phonetic alphabets to the printing press, telegraph, radio, and television, has reopened social and political debates about how the content of public communication should be regulated. Book-banning has a long and infamous history, in both democratic and nondemocratic traditions, as does wartime censorship of the press. During the McCarthyite 1950s, many suspected "reds" were blackballed from Hollywood, radio, and the print media. TV censors fretted about showing Elvis's hips on the Ed Sullivan show; hillbilly rock 'n' roll, it seems, was too sensual for polite company. Today concern abounds over 1-900 telephone sex services, the kinds of advertising that should be allowed on television (ranging from condom ads to toy ads during Saturday morning cartoon broadcasts), and cable pornography channels. In the realm of politics, debate continues about the kinds of programming that should be allowed on public television stations and public access cable channels. For many Americans, it seems, open access to information is too risky; the implication is that viewers are easily manipulated by TV shows and advertising—that consumer sovereignty fails when certain kinds of anti-establishment ideas are expressed in public.

Television is not the latest but is perhaps the greatest of these contested arenas of information transfer.[2] Today we have not just television but sixty-nine or more channels of computer-enhanced, tape-recorded, stereo-capable, picture-in-a-picture TV. And, yes, it has been blamed for everything from illiteracy to the crumbling of sexual mores in America.

Some of the apparent ills of political competition noted in the preceding essays do seem plausibly attributable to TV. Sabato's argument, for example, is fairly compelling: (1) TV networks created competition in national news coverage; (2) Watergate showed that it was possible to create marketable, national news stars, thereby changing incentives for print and TV journalists and also changing the expectations of the people who pay them. According to Sabato, the increasing resources expended by the networks and the accompanying increased competition, together with the change in journalists' incentives created by the star system, led to an open season of pack journalism directed toward presidential contenders.

More important, the decline of party machines and the ascendancy of TV are alleged to have engendered further changes that many claim are shaking the foundations of our democracy. A number of fairly specific conclusions have been drawn from the observations discussed above. First, poor candidates have been chosen, not only in terms of the representativeness of their policy views[3] and their responsiveness to voters' preferences[4] but also their competence.[5] Second, because bad candidates are nominated, it follows that poor presidents are chosen, as a result of which both policy choices and their execution are inferior.

One alleged indication of this is the increasing failure of presidents to get along with Congress.[6] Another indication is an observed decline in collective responsibility for policy choices and their execution. Instead of writing specific legislation and fixing problems, modern American politics is often said to consist largely of finger-pointing, speech-making, position taking, and blame avoidance.[7] Ultimately, the result is paralysis, or, in the event that a policy is chosen, the effect is suboptimal or contrary to central tendencies in voter preferences.[8]

Third, because presidents, policy choices, and governance are all inferior, voters' dissatisfaction with government has increased and, consequently, voter participation has decreased. This argument is commonly made by newspaper columnists across the country and is echoed in a number of recent studies of public opinion.[9] Critics charge that these three observations lead to the conclusion that American government is losing its legitimacy as a democratic institution. As David Broder argues, the decline of political machines

> may in the long run contribute to the decay of political legitimacy within the constitutional order. This may be reflected in the palpable growth of disaffection with Presidents and in the phenomenon frequently complained of—mostly by neoconservative observers—and labeled "ungovernability," a pervasive inability of political leaders to satisfy the expectations of voters.[10]

Seemingly longing for the days of smoke-filled rooms and minimal popular input into the selection process, many political observers excoriate the modern process of nominations through popular primaries for replacing issue content and "real" politics with opiates for the masses—sex scandals, horse-race press coverage, and, in general, anything but serious treatment of the issues.[11] The change to the Australian secret ballot in the late nineteenth century, the introduction of primaries in the early twentieth century, and the reforms of the McGovern-Fraser and later commissions have obviously produced winners and losers in the two parties' internal processes. I am not surprised that the losers complain bitterly and liken these losses to the decline of American democracy itself.

Telling Wheat from Chaff

The above are the allegations. On observing an alleged fact we can explain it or explain why people think it is a fact. Often the latter exercise is more interesting, especially when there is no evidence the allegation is true, as is the case for the alleged poor presidents. In this section I attempt to explain, in two parts, why people have come to believe these allegations. First, I argue that people's thinking about political parties has not kept pace with what these have become. Thus, the perceived decline of party is really a change in what party is and what it does. Second, people have misunderstood the institutional role of the president in our constitutional system and, as a consequence, have blamed problems arising from divided government, which has existed for most of the postwar period, on problems with presidents; they can't get along with Congress, they don't try to work with Congress, etc. I consider these two points in turn.

Party Time

The standard academic views of political parties fall into two broad categories, neatly explicated by Wattenberg in *The Decline of American Political Parties*: studies of organized groups that control the machinery of government to make public policy and studies of political opinions and attitudes in the electorate.[12] My primary concern in this section is the former approach—parties as policy-making organizations. Scholars, at least from the time of Robert Michels,[13] have thought of parties in terms of an ideal type of organization: centralized, hierarchically organized, tightly disciplined—in a word, a political machine.

Critics rightly note that legislative parties in general are not homogeneous, but it does not follow that party members will be unable to cooperate consistently with one another on some set of policy questions. To the extent that such cooperation persists and is observable to voters, party labels provide valuable signals about the likely future behavior of a candidate for legislative office. If office seekers expend resources and effort to win party nominations, voters can reasonably infer that the party label is valuable to candidates, and hence that incumbents are likely to have incentives to protect the electoral value of that label.

Thus, although the era of strong political machines is said to have passed into history, replaced by the age of candidate-centered elections, competition for the right to bear a major party label in a general election remains relevant. Perhaps an even stronger argument in favor of the label's value is derived from juxtaposing rising expenditures in primary campaigns with survey research indicating that fewer Americans now identify themselves as "strong" partisans. If we interpret the survey findings to mean that voters care less and less about party labels, why would candidates be willing to spend more to gain nomination?

The point is that party-as-machine is the wrong way to think about parties in American politics. Machine organizations have at different times run many different governments in many different places in the United States, but their existence is not a necessary condition for the party label to have value to voters, nor is machine control of a government a necessary condition for partisan cooperation in setting policies or managing delegations to bureaucrats.

As an illustration of this point, think about the behavior of franchise holders in the marketplace, such as owner-operators of McDonald's restaurants. The McDonald's corporation sells the right to use its brand name and to distribute the paraphernalia that goes with its products: the characteristic foods and golden arches-inscribed cups, bags, etc. McDonald's expends effort to monitor the behavior of franchise-holders, but there are many thousands of individual restaurants around the world, suggesting that direct oversight would likely be a very costly approach to protecting the brand name. Instead, McDonald's depends to a great extent on the individual interests of its owner-operators: since consumers know the brand name so well, they expect the same quality service, food, and ambience at every McDonald's they patronize. Any restaurant that fails to provide any of the expected visual signals, before the customer places an order, will raise doubts as to whether the food will conform to McDonald's standards—will put the brand name into question—and hence is likely to drive away customers.

The same logic applies to presidential nominating campaigns. Candidates are, in effect, trying to gain the right to use a brand name (the party label) for the general election. Nothing, in fact, prohibits the losers from running in November without the party label. Indeed, losers in the nomination process do sometimes run on third-party labels: George Wallace in 1968 and John Anderson in 1980 are two prominent examples. But no third-party candidate since Theodore Roosevelt in 1912 has come close to winning the general election. It seems clear that winning the nomination of one of the major parties is still a necessary condition for winning the presidency.

Candidates for party nomination are not all born equal. Some have more established partisan track records or reputations than others; voters can reasonably infer from a long record of partisanship that a Republican will remain a Republican once in office, and likewise a Democrat will remain a Democrat. Indeed, candidates always seem to dedicate a significant part of their nomination campaigns to advertising their track records and previous service in order to convince voters in primaries and caucuses that they are the genuine article—one of "us," not one of "them." Another way for a candidate to get a leg up on winning nomination is to collect endorsements from individuals and groups with established reputations, such as businesses, labor unions, celebrities, incumbent officeholders and political machines.

Of course, machine organizations did seem to be more prevalent during the late nineteenth and early twentieth centuries than they are today. A machine is two things. First, it is an organization with an established reputation (vis-à-vis its preferred public policies) that provides public endorsements for candidates. Second, it is a cartel of people committed to voting as a bloc. As a rule of thumb, machines tend to be partisan: there have been Republican machines and Democratic machines, but no significant examples of bipartisan machines. A simple inference from this observation, then, is that association with a machine provides one way for candidates to show commitment to a party label.

Machines also deliver blocs of votes to favored candidates. If no machine is capable of delivering a winning share of votes, candidates will have to look beyond machine support to secure office. Many of the electoral reforms of the last century, starting with the institution of the Australian ballot, have made it more difficult for machines to maintain large voter cartels. Television has played an important role in this process by lowering the cost to voters of collecting information about alternate candidates, thus reducing the relative importance to candidates of endorsements from political machines. It does not follow, however, that party labels have declined in importance. On the contrary, much of what today constitutes personal reputation building consists of candidates showing how their track records conform to public expectations of partisan behavior in office.[14] Now, more than ever, it is up to the candidate to show he or she is a good Republican or Democrat and not a loose cannon.

Presidents and Power

If the decline of partisanship and party machines doesn't explain why so many scholars believe that nomination processes produce poor candidates and presidents, what does? The answer lies in how scholars have come to think about the presidential office itself. Models of presidential dominance—that the president sets the legislative agenda, dominates the budgetary process, controls appointments, etc.—have proliferated in academia during the postwar period. I argue that people have taken the claims of the presidential dominance approach at face value while simultaneously observing various failures of governability, especially since Vietnam. Faced with a conflict between theory and observed outcomes, scholars and casual observers have chosen to punt, arguing that the theory of presidential power is right, but that the presidential selection process has failed to produce competent, centrist candidates.

The presidential dominance thesis argues that Congress in the twentieth century has been unable to resist executive encroachments on its constitutional prerogatives. It has become a stylized fact that the power of the executive has outstripped the ability of the other two branches to fully check it.[15] Of course, most of the perceived shift has not been the result of encroachment per se, but rather through delegations of authority by act of Congress. In one important policy area after another, Congress is said to have delegated its authority to the executive and then turned its back on the results of that delegation, i.e., abdicated responsibility. An oft-cited example is the Budget and Accounting Act of 1921, which defined the basic framework for the executive budget process and created the Bureau of the Budget (now Office of Management and Budget).

The thesis of presidential dominance fails to explain the considerable lengths to which members of Congress go to delegate their authority to the executive, as recent books by Aberbach and by Kiewiet and McCubbins have shown.[16] Indeed, I have argued elsewhere that there is very little evidence to support any of the general claims of presidential power outside of the influence conferred on the president by the constitutional veto power—and, as Kiewiet and McCubbins have shown, the influence conferred by the veto power is both limited in scope and asymmetric.[17]

But the veto power clearly gives the president a stake in the policy making process; in other words, voters have good reason to evaluate incumbent presidents based on actual policy outcomes and nonincumbent candidates based on expectations about their marginal effects on policy outcomes. The question is: can voters make accurate inferences about incumbents' marginal contributions to outcomes and forecasts of each candidate's likely impact on future policies? The presidential dominance approach is a dead end: it provides very little help in understanding past outcomes or forecasting future ones. A different approach is needed—one in which the institutional authorities and limitations of the president are taken seriously.

Two recent approaches in the literature offer some hope of understanding how voters evaluate the presidential office. I have argued from an institutional perspective that divided government leads to quite predictable outcomes. When a president from one party faces a unified partisan majority of the other party in Congress, as during the Nixon, Ford, and Bush administrations, the president's party is generally in a disadvantageous position for implementing its preferred policies. Congress controls the legislative agenda, both in terms of how bills are packaged and when they are offered to the president for consideration. The president has no institutional means through which to force Congress to provide the desired result. Thus it may be possible for Democratic Congresses to, in effect, buy off Republican presidents with small concessions buried within otherwise solidly Democratic bills.

Under these circumstances, how should a president be evaluated by voters? Most presidential accomplishments under conditions of divided government tend to be negative, such as Republican pledges of "no new taxes." As a practical matter, such negative achievements are hard to sell to constituents looking for reasons to retain an incumbent.

As for positive activities, Republican presidents in the postwar period have tended to emphasize their foreign policy accomplishments. This is the one area of policy in which the president can be said to possess some agenda-setting power, since the Constitution specifically delegates authority to the president to negotiate treaties and serve as commander in chief of the armed forces. The structural difference between foreign and domestic policy arenas suggests the second approach to understanding voter evaluations of presidential candidates. Jacobson has argued that voters want and expect different things from presidents than from Congress:

> People want mutually exclusive things from government. There is nothing irrational here; we naturally enjoy the benefits government confers but dislike paying for them. . . .[18]

> Perceived differences between the parties coincide with differences in what people expect of presidents and members of Congress, which in turn reflect differences in the political incentives created by their respective institutional positions. Presidents are supposed to pursue broad national interests; uniquely among elected officials, presidents can profit politically by producing diffuse collective benefits at the expense of concentrated particular interests.[19]

Jacobson suggests that voters use different criteria to evaluate presidential and congressional candidates. He argues that these different criteria arise from the fact that the president faces a national constituency, whereas members of Congress face narrow, localized constituencies. In other words, voters implicitly or explicitly recognize that members of Congress find it difficult to act collectively, whereas the president fully internalizes the conflicting interests of different groups of voters when making policy choices.[20]

For strong partisan identifiers, this sort of institutional specialization poses no problems for vote choice; however, for those voters less firmly attached to one party or the other, such specialization seems likely to induce splitting their tickets, without requiring the performance of mental gymnastics to compose a sophisticated voting strategy to get the outcomes they want. If the two parties are fairly evenly matched in terms of numbers of strong identifiers, but the weak identifiers are increasingly splitting their tickets in conformity to Jacobson's model, it follows that electoral results will increasingly reflect the ticket splitting of the weak identifiers. It also follows that both groups of strong identifiers will find themselves increasingly unable to do anything about these results, which both groups find distasteful, and hence we may expect to see increasing dissatisfaction with divided government on the part of two large segments of the electorate—those who would like unified Republican government and their counterparts on the Democratic side.

Two possible explanations for recent scholarly dissatisfaction with the quality of presidents are available, both of which ultimately highlight divided government as the root cause of policy choices reviled by partisans at both ends of the ideological spectrum. I emphasize the gulf between many scholars' expectations about presidential power and the theoretical results that emphasize the structural limitations and asymmetric role embedded in the presidential veto; Jacobson's argument, on the other hand, would seem to suggest that scholars have failed to understand that voters actually want and expect different things from the president than from Congress. He maintains that when voters separate their presidential choices from congressional choices, the electoral consequences are a higher probability of divided government. On the one hand, then, scholars are frustrated that their imagined philosopher-king presidents fail to dominate Congress; on the other, they see frustrated strong partisan voters dropping out and weak identifiers splitting their tickets, from which scholars have mistakenly inferred that party labels no longer mean anything.

The purpose of this volume has been to present in detail the substance of the scholarly beefs about presidential campaigns in order to separate fact from fiction. I have argued vociferously that almost none of the barking dogs in this literature have any teeth. Larry Sabato has shown that news coverage of presidential campaigns has changed considerably since the advent of television—and particularly since Watergate opened the door to celebrity journalism. The consequences for presidential selection, however, are unclear. Vituperative reporting may be in bad taste, but it does not obviously bias the field of candidates other than to knock out those with particularly thin skins or skeleton-laden closets.

Aldrich points out that television has increased the velocity with which political information spreads, making it much more difficult for candidates to manage changes in their reputations. He suggests that this partially accounts for the rise of candidate-centered campaign organizations: media consultants, public relations experts, etc. But the consequences for presidential selection are not altogether clear. Candidates still vie for party nominations and still must build personal reputations based on past service, usually as partisan officeholders. If television has had any effect, it seems most likely to have increased the relative chances of good partisans and experienced politicians over less well-known or weaker partisans. Thus, argue Arterton and Popkin, campaigns and campaign advertising provide information that helps voters to make better choices. Arterton further argues that campaign advertising is selective in the kinds of information it tries to convey to voters. Specifically, campaign ads send signals to voters about their sponsor's good attributes and the bad attributes of other candidates. And, he argues, negative ads are more likely to affect voters' decisions than are positive ads.

There is very little evidence to support this general claim, but it does seem to contain an element of truth, namely, that competitors for elective office potentially hold each other's reputation hostage during a campaign. Each competitor has an incentive to exploit any relative weakness of other candidates in order to give himself or herself an advantage over the field; for example, if candidate A knows or believes that candidate B has used drugs, A will have an incentive to reveal that information to reporters. Reporters, in turn, have an incentive to publicize the information, thus providing, in effect, free advertising to candidate A. But this story is not about paid media; rather, it relies on free media. Indeed, as far as paid advertising is concerned, there should be little or no difference between the effects of negative and positive ads.

Finally, Popkin argues that campaigns, rather than being too long, are still too short. The longer the campaign, the more signals the successful candidate can send to voters. Presidential campaigns before television, Popkin suggests, were much different from those of today because candidates were so closely tied to party machine organizations and could make credible commitments to policy positions precisely because of these ties. In essence, voters didn't need to know every aspect of a candidate's preferences because they knew the reputation of the machine providing the candidate's support. Today's candidates, however, generally do not have the option of relying solely on boss endorsements; they must actively court voters. If there is competition for votes, candidates will have incentives to continue revealing information until voters believe that they are able to make the same choice of candidate that they would have had they acquired complete encyclopedic information.

The argument of this book is twofold: first, the essayists and I argue that presidential campaigns are important political phenomena, not merely placebos or bad entertainment. We differ, however, on how that observation is to be applied. Some of the authors contend that the changes in presidential campaigns have made American democracy more unstable by alienating voters, or at least making it more difficult for them to determine which candidate best satisfies their interests. In contrast, I argue that, although the relative decline of party machines and the rise of television have changed how campaign resources are expended and campaign organizations are structured, they have not changed either the quality or the ideological positioning of successful candidates. If you the reader have come away better able to evaluate the relative merits of these and similar arguments about political choices and institutions, this volume will have been a success.

Notes

1. Joseph Schumpeter, *Capitalism, Socialism and Democracy* (New York: Harpers, 1942).
2. Subsequent to television, of course, are such technologies as satellite communication, fax machines, and electronic (computer) bulletin boards.
3. For example, it is widely held that Goldwater in 1964 and McGovern in 1972 were outlier candidates, even relative to the central tendency of preferences in their own parties. See, e.g., Gary Jacobson, *The Politics of Congressional Elections*, 3rd ed. (New York: HarperCollins, 1992), 110-111.
4. See, e.g., Lance Bennett, *The Governing Crisis* (New York: St. Martin's Press, 1992), 28.
5. The latter claim is implicit in the arguments of many authors, including Nelson W. Polsby, *Consequences of Party Reform* (Oxford: Oxford University Press, 1983), and William J. Crotty, *Decisions for the Democrats: Reforming the Party Structure* (Baltimore: Johns Hopkins University Press, 1978).
6. Polsby, *Consequences of Party Reform, 114;* Nelson W. Polsby and Aaron Wildavsky, *Presidential Elections: Contemporary Strategies of American Electoral Politics,* 7th ed. (New York: Free Press, 1988), 39; Austin Ranney, "The Democratic Party's Delegate Selection Reforms, 1968-76," in *America in the Seventies,* ed. Allan P. Sindler (Boston: Little, Brown, 1977), 204.
7. Morris Fiorina, *Congress: Keystone of the Washington Establishment,* 3rd ed. (New Haven, Conn.: Yale University Press, 1990); Theodore Lowi, *The End of Liberalism,* 2nd ed. (New York: Norton, 1979).
8. David Broder, *The Party's Over* (New York: Harper & Row, 1972); Bennett, *The Governing Crisis.*
9. Howard L. Reiter, "The Limitations of Reform: Changes in the Presidential Nominating Process," Essex Papers in Politics and Government, No. 20 (Department of Government, University of Essex, Wivenhoe Park, Colchester, England, 1984): 1.
10. Broder, *The Party's Over,* 141-142.
11. See, e.g., Robert E. DiClerico and Eric M. Uslaner, *Few Are Chosen:*

Problems in Presidential Selection (New York: McGraw-Hill, 1984), and John H. Aldrich, *Before the Convention: Strategies and Choices in Presidential Nomination Campaigns* (Chicago, Ill.: University of Chicago Press, 1980).

12. Martin Wattenberg, *The Decline of American Political Parties, 1952-1984* (Cambridge, Mass.: Harvard University Press, 1986), 3-4.

13. Robert Michels, *Political Parties*, Eden Paul and Cedar Paul, trans. (New York: Free Press, 1958). Michels's book, which first appeared in 1915, argued that the delegation of authority in organizations from rank and file to leaders results in oligarchy or dictatorship within the group—his "iron law of oligarchy."

14. Recall Sen. Robert Dole's parting shot at then Vice President George Bush during a New Hampshire debate: "Stop lying about my record!"

15. This literature is vast. See, e.g., Wilfred E. Binkley, *President and Congress*, 3rd ed. (New York: Vintage Books, 1962); James Bryce, *The American Commonwealth* (New York: Macmillan, 1924); and Richard E. Neustadt, *Presidential Power: The Politics of Leadership from FDR to Carter* (New York: Wiley, 1980).

16. Joel Aberbach, *Keeping a Watchful Eye* (Washington, D.C.: Brookings Institution, 1990); D. Roderick Kiewiet and Mathew D. McCubbins, *The Logic of Delegation* (Chicago, Ill.: University of Chicago Press, 1991).

17. D. Roderick Kiewiet and Mathew D. McCubbins, "Presidential Influence on Congressional Appropriations Decisions," *American Journal of Political Science* 32 (1988): 713-736. The paper argues that presidents can use the threat of a veto to pull down appropriations when the president wants to spend less on a program than do majorities in Congress, but he cannot extract higher appropriations when he wants to spend more.

18. Gary C. Jacobson, *The Electoral Origins of Divided Government: Competition in U.S. House Elections, 1946-1988* (Boulder, Colo.: Westview, 1990), 106.

19. Jacobson, *Divided Government*, 112.

20. A similar argument has recently been made elsewhere on trade policy by Susanne Lohmann and Sharyn O'Halloran ("Delegation and Accommodation in U.S. Trade Policy," unpublished paper, Stanford University, November 1991); and on tariff policy making by Judith Goldstein and Barry Weingast ("The Origins of American Trade Policy: Rules, Coalitions and International Politics," paper prepared for the Social Science Research Council Conference on Congress and Foreign Policy, Stanford University, April 21-22, 1991). These two papers are explicitly about policy making and the delegation of authority by Congress to the president. Lohmann and O'Halloran argue that members of Congress delegate to the president on aspects of trade policy and precommit to restrictive rules for consideration of presidential proposals, such as fast-track proceedings, to avoid the inefficiencies that would arise in logrolling

negotiations over the same policies if Congress were to do the work itself. Goldstein and Weingast argue similarly that Congress delegated authority to the president to negotiate bilateral trade agreements that lower tariffs on specific goods to take advantage of the president's preference for policies favoring broad national constituencies over narrow particularistic ones.

CONTRIBUTORS

John H. Aldrich is professor of political science at Duke University. He received his Ph.D. from the University of Rochester. He is author of *Before the Convention: Strategies and Choices in Presidential Nomination Campaigns* (1980) and coauthor of *Change and Continuity in the 1980 Elections* (CQ Press, 1982, 1983), *Change and Continuity in the 1984 Elections* (CQ Press, 1986, 1987), and *Change and Continuity in the 1988 Elections* (CQ Press, 1990, 1991). His articles have appeared in *American Political Science Review, American Journal of Political Science, Public Choice, Journal of Politics,* and *Social Psychology Quarterly.* He currently is serving as chairman of the Department of Political Science at Duke.

F. Christopher Arterton is the dean of The Graduate School of Political Management. He received his Ph.D. from the Massachusetts Institute of Technology. He is author of *Teledemocracy: Can Technology Protect Democracy?* (1987) and *Media Politics: The New Strategies of Presidential Campaigns* (1984) and coauthor of *The Electronic Commonwealth: The Impact of New Media Technologies on Democratic Politics* (1988) and *Explorations in Convention Decision Making* (1976). Arterton has been a polling consultant for numerous Democratic campaigns at the local, congressional, and gubernatorial levels. Since 1979 he has served as a consultant for the *Newsweek* poll. Arterton has been elected delegate to the Democratic National Convention and was a member of several national commissions on party rules for the Democratic National Committee.

Mathew D. McCubbins is professor of political science at the University of California, San Diego. He received his Ph.D. from the California Institute of Technology. McCubbins is coauthor of *The Logic of Delegation* (1991) and *Legislative Leviathan* (forthcoming) and coeditor of *Congress: Structure and Policy* (1987). He has published numerous articles in the *American Journal of Political Science, Legislative Studies Quarterly, Public Choice, Journal of Poli-*

tics, *Journal of Law, Economics and Organization, Virginia Law Review, Journal of Institutional and Theoretical Economics,* and many other social science and legal journals.

Samuel L. Popkin is professor of political science at the University of California, San Diego. He received his Ph.D. from the Massachusetts Institute of Technology. He is author of *The Rational Peasant: The Political Economy of Peasant Society in Vietnam* (1979), coauthor of *Candidates, Issues, and Strategies* (1965), and coeditor of *Chief of Staff: Twenty-five Years of Managing the Presidency* (1986). Popkin has served as a consultant on polling and strategy to the George McGovern, Jimmy Carter, and Bill Clinton presidential campaigns. He also has assisted with polling and election coverage for CBS News.

Larry J. Sabato is the Robert Kent Gooch Professor of Government and Foreign Affairs at the University of Virginia. He is a former Rhodes Scholar and Danforth Fellow and received his B.A. from the University of Virginia and his doctorate from Queen's College, Oxford University. His books include *The Rise of Political Consultants: New Ways of Winning Elections* (1981), *Goodbye to Good-time Charlie: The American Governorship Transformed* (CQ Press, 1983), *PAC Power: Inside the World of Political Action Committees* (1984), *The Party's Just Begun: Shaping Political Parties for America's Future* (1988), and *Feeding Frenzy: How Attack Journalism Has Transformed American Politics* (1991).

INDEX